# A MANAGER'S GUIDE TO PRODUCTIVITY, QUALITY CIRCLES, AND INDUSTRIAL ROBOTS

**Harry Katzan, Jr.**

**VNR** VAN NOSTRAND REINHOLD COMPANY
—————————————————— New York

Copyright © 1985 by Van Nostrand Reinhold Company Inc.

Library of Congress Catalog Card Number: 84-25739
ISBN: 0-442-24823-7

All rights reserved. No part of this work covered by the copyrights hereon may be reproduced or used in any form or by any means—graphic, electronic, or mechanical, including photocopying, recording, taping, or information storage and retrieval systems—without permission of the publisher.

Manufactured in the United States of America

Published by Van Nostrand Reinhold Company Inc.
135 West 50th Street
New York, New York 10020

Van Nostrand Reinhold Company Limited
Molly Millars Lane
Wokingham, Berkshire RG11 2PY, England

Van Nostrand Reinhold
480 Latrobe Street
Melbourne, Victoria 3000, Australia

Macmillan of Canada
Division of Gage Publishing Limited
164 Commander Boulevard
Agincourt, Ontario MIS 3C7, Canada

15  14  13  12  11  10  9  8  7  6  5  4  3  2  1

**Library of Congress Cataloging in Publication Data**
Katzan, Harry.
  A manager's guide to productivity, quality circles, and industrial robots.

  Bibliography: p.
  Includes index.
  1. Industrial productivity.  2. Quality circles.
3. Robots, Industrial.  I. Title.
HD56.K377   1985      658.3'14         84-25739
ISBN 0-442-24823-7

# PREFACE

Japanese business is in the news these days, because of its high productivity. This success is largely the result of two important factors: quality control and industrial robotics. Although Japanese culture and management style are unique, the underlying technology is not. In fact, most of it was developed in the United States. As a result, there is a considerable amount of U.S. and European interest in three relevant subjects:

- Productivity
- Quality circles
- Industrial robots

In general, the topic of productivity provides a shell for the other subjects, which deal with people and machines, respectively.

The book is divided into three main sections:

- Productivity and quality
- Quality circles
- Industrial robots

and seven chapters. Topics covered in the section on productivity and quality include productivity as a concept, work, technology and employment, economics and productivity, jobs, productivity improvement, the concept of quality, the quality life cycle, quality assurance and control, quality awareness, and related organizational issues.

The section on quality circles covers the concept of participative management from policy formulation to the dynamics of quality circle operation. Methods and planning for quality circles are also covered in the treatment.

The section on industrial robots covers basic and advanced concepts in robot design and applications. Robot anatomy, control, geometry, operation, development, and programming are covered in a chapter on robot principles; and flexible manufacturing systems, machine vision systems, and CAD/CAM are covered in a section on advanced concepts.

Four basic definitions are relevant to the subject matter of the book:

- *Productivity* is defined as the relationship of the output of goods and services to the input of labor and capital to the total process.
- *Quality* is defined as the totality of attributes of a product or service that reflect on its innate capacity to satisfy a given set of needs.
- A *quality circle* is a small group of workers who meet regularly on a voluntary basis to analyze problems and recommend solutions to management.
- A *robot* is a reprogrammable, multifunctional manipulator designed to move material, parts, tools, or specialized devices through variable programmed motions for the performance of a variety of tasks.

As such, the four topics collectively serve as a basis for decision, control, and action in all types of organizations of all sizes. The subject matter is as applicable to government and service industries as it is to manufacturing, banking, and other forms of commerce.

This book is intended for executives, managers, administrators, analysts, and scientific and technical people. No special background other than an interest in productivity and quality is required.

It is a pleasure to acknowledge the assistance of my wife, Margaret, who helped with manuscript preparation and was a good partner during the entire project.

Harry Katzan, Jr.<br>
Millstone Township, New Jersey

# CONTENTS

**Preface / iii**

**Part One PRODUCTIVITY AND QUALITY**

## 1. PRODUCTIVITY CONCEPTS / 3

INTRODUCTION / 3
PRODUCTIVITY AS A CONCEPT / 4
IMPORTANCE OF PRODUCTIVITY / 5
WORK / 9
TECHNOLOGY AND EMPLOYMENT / 10
ECONOMICS AND PRODUCTIVITY / 12
PRODUCTIVITY, QUALITY, AND JOBS / 13
BASIC INGREDIENTS IN A PRODUCTIVITY IMPROVEMENT
   PROGRAM / 13
SUMMARY / 14

## 2. QUALITY CONCEPTS / 16

INTRODUCTION / 16
CONCEPT OF QUALITY / 17
QUALITY LIFE CYCLE / 18
QUALITY ASSURANCE / 21
METHODS OF QUALITY CONTROL / 22
QUALITY AWARENESS / 25
QUALITY POLICY, OBJECTIVES, AND PLANNING / 27
QUALITY ORGANIZATION / 29
SUMMARY / 30

**Part Two QUALITY CIRCLES**

## 3. QUALITY CIRCLE PRINCIPLES / 35

INTRODUCTION / 35
ORGANIZATION FOR QUALITY CIRCLES / 37
POLICY FORMULATION / 39
IMPLEMENTATION / 44

QUALITY CIRCLE PARTICIPANTS / 44
  Steering Committee / 47
  Facilitator / 49
  Circle Leader / 50
  Circle Members / 52
OPERATION OF A QUALITY CIRCLE / 53
VARIETIES OF QUALITY CIRCLES / 55
SUCCESS ELEMENTS / 56
SUMMARY / 57

## 4. QUALITY CIRCLE METHODS / 60

INTRODUCTION / 60
MEASUREMENT / 60
  Quality Indicators / 61
  Cost Analysis / 62
  Attitude Indicators / 64
BRAINSTORMING / 65
  Basic Concept / 65
  Guidelines / 66
  Analysis of Ideas / 67
DATA COLLECTION / 67
  Recording of Collected Data / 68
  Presentation of Collected Data / 70
  Advanced Methods / 73
  Discussion / 74
DATA ANALYSIS / 74
  Pareto Diagrams / 74
  Basic Cause and Effect / 77
  Process Cause and Effect / 77
APPLICATIVES / 79
SUMMARY / 80

## 5. STRATEGIC PLANNING FOR QUALITY CIRCLES / 82

INTRODUCTION / 82
SYSTEMS APPROACH TO PLANNING / 82
STRATEGY CONCEPTS / 84
STRATEGY AND PLANNING / 86
  Contents of the Strategy / 86
  Strategic Plan / 87
  Guidelines / 88
JUSTIFICATION / 88
  Definition / 88
  Methodology / 89

Productivity / 89
   Methods of Analysis / 89
IMPLEMENTATION / 90
   The Pilot Project Approach / 90
   Selecting a Climate for Success / 91
   Choosing the Right Shop or Department / 91
EMPLOYEE ACCEPTANCE / 91
   Acceptance Strategy / 92
   Resistance Management / 92
   Education and Training / 93
SUMMARY / 93

**Part Three INDUSTRIAL ROBOTS**

**6. ROBOT PRINCIPLES / 99**

   INTRODUCTION / 99
   ROBOTICS SCENARIO / 100
   ROBOT CONCEPTS / 104
   ROBOT ANATOMY / 105
   ROBOT GEOMETRY / 106
   ROBOT DEVELOPMENT / 108
      Remote Manipulators / 109
      Numerical Control / 110
      Pick-and-Place Robots / 110
      Programmable Robots / 111
   ROBOT CONTROL / 112
      Actuators / 112
      Arm Control / 113
      Memory / 114
   ROBOT PROGRAMMING / 114
   SUMMARY / 116

**7. ADVANCED TOPICS IN ROBOTICS / 119**

   INTRODUCTION / 119
   FLEXIBLE MANUFACTURING SYSTEMS / 120
   MACHINE VISION SYSTEMS / 122
   CAD/CAM AND ROBOTICS / 124
   SUMMARY / 126

**References / 129**

**Index / 133**

# PART ONE
# PRODUCTIVITY AND QUALITY

# 1 PRODUCTIVITY CONCEPTS

**INTRODUCTION**

Productivity is in the news these days as an effective means of counteracting the downward effects of various other important economic forces, but only in the sense that more of a good should in general be preferred to less of a good. In fact, it has often been said in business circles that if less of a good were preferred, it would instead be called a "bad." The consideration is not to be taken lightly, since a lower productivity in some areas can be balanced with a higher productivity in other domains. An illustrative example of where lower worker productivity could be desirable exists in the construction industry where premolded bathrooms or prefabricated houses are installed. The productivity of installation workers in this area is generally regarded as being relatively low because the requisite variety of the task is uncommonly high. Yet, the "total productivity" resulting from the use of this technology through subcontracting of this type is high because of the specialization and economy of scale inherent in the assembly, fabrication, and molding processes. The key point is that productivity regarded as per-person output, as is frequently the case, is best analyzed from the standpoint of the enterprise as a whole.

Another important consideration is the notion that the only aspect of productivity worth considering is worker output, and that "job satisfaction" in many of its derivative forms is the key issue. Roughly only 10 to 25 percent of total productivity is worker dependent. Clearly, the reasons for the belief that worker productivity is paramount are largely historical, but they nevertheless exist and must be dealt with. The worker remains the key element in a production system, but social scientists now realize that an emphasis on job satisfaction is an oversimplification of the total problem.

A final consideration is the belief that a worker can be replaced by a machine—either intelligent, such as a robot, or nonintelligent, such as a special-purpose device—and that is all there is to productivity. Automation—if in fact it is a good idea at all, and many of us now feel that it is—requires a balanced approach that necessitates a very close look at the entire system and definitely not a brief perusal of a few operations in a complete production process.

Each of these topics is covered to varying degrees of detail in this book. The issues are organizational, social, and technical, and as such, require unique avenues of exposition.

## PRODUCTIVITY AS A CONCEPT

The notion of productivity is rooted in history, but not as far back as most people think. Before industrialization, there was little need to even consider the subject of productivity. Without commerce or storage facilities of various kinds, the incentive for a farmer to produce more than the immediate family could consume was practically nonexistent. Through primitive forms of bartering, some excess crops could be traded for hard goods. Otherwise, there was obviously no need for a higher level of production than could be comfortably achieved. Actually, this mode of behavior is a direct carryover from feudal periods, when additional crops were taken directly by an overseer— often a king or ruling party.

With the introduction of farm implements, mechanization was largely based on personal convenience rather than increased production. In subsequent cultures, starting with ancient Greece and the Roman Empire and extending through periods of slavery and early industrialization (i.e., the early "sweatshops"), increased production from the viewpoint of the individual served only as a means of avoiding punishment in a variety of well-known forms.

With the advent of money as a medium of exchange, together with a sufficient level of industrialization, the output of the production process became of prime importance. The level of input was largely ignored as long as output continued to increase. This naturally lead to division of labor and specialization aided physically by machines of various kinds to enhance output. Rudimentary considerations of productivity effectively started during this period.

*Productivity* is defined in real quantities as the relationship of the output of goods and services to the input of labor and capital to the total process. While the term "productivity" usually connotes a production or manufacturing operation, the concept applies equally well to the service industries and other forms of commerce, such as banking and retailing. Productivity measures are commonly applied to quantities that can be easily counted. Typical input values are hours worked, tons of steel, machine hours utilized, and money consumed. Typical output values are units produced, applications accepted or rejected, persons educated, cases resolved, units sold, and so forth.

Recent productivity activities designed to increase worker performance center on four areas of interest:

- Worker satisfaction
- Job redesign
- Motivation and attitude
- Direct employee incentives

As would ordinarily be expected, each of these areas has importance to the modern organization and clearly should not be neglected. However, it is cumbersome and tedious to dispel the belief that productivity is not solely a function of labor for both historical and technical reasons. In the latter case, it is simply more convenient to concentrate on worker behavior—even though overall gains may be relatively small. New technology requires both capital and commitment, and in many enterprises, everyday operational concerns simply supersede the advanced planning necessary for the introduction of new facilities. Many managers are simply more competent at and comfortable with worker considerations than with technical issues.

**IMPORTANCE OF PRODUCTIVITY**

In many ways, productivity is associated with division of labor and specialization. In the professions, this concept is desirable. In the factory, it is commonly regarded as dehumanizing. Professional specialists generally demand higher wage levels, whereas worker specialists frequently complain that their work is tedious and monotonous. Through unionization, workers have been awarded relatively high levels of wages when the factors of education and experience, as compared to professionals, are duly considered. Worker specialization is not necessarily a direct result of unionization, but the association is clearly evident.

The elements of division of labor and specialization are not necessarily undesirable from a productive point of view. In fact, it is precisely these elements that make a direct contribution to productivity—at least initially—because they pave the way for three important advantages:

- A higher level of individual worker skill
- Increased work flow organization
- The use of machines to augment the worker

Productivity elements such as these have enhanced the competitiveness of the work environment. At first, the guise was patriotic duty: enabling one's country to compete effectively in the international marketplace. Thus, workers were encouraged to be more productive because it provided a means for the country to price its products below the prices of competing countries. Lately, however, the guise for increased productivity has been more desperate: survival. Many business enterprises find it increasingly difficult to compete because their competitors have become *more* than productive people; they have a more productive system, which is essentially a means of "working smarter."

A *productive system* is one that contains five basic ingredients:

- Necessary resources
- Optimum proportion of resources
- Maximum utilization of each resource
- A productive environment
- Proper interaction between resources

With these elements combined in an effective manner, an operating structure is achieved that is appropriate for attaining the "success factors" of the organization. The success factors can of course be units, heads, rejects, or whatever countable measures of productivity exist. The production process combines elements of the five ingredients— people, machines, input goods, etc.—in the optimum proportion permitting each to be best utilized while fully interacting

PRODUCTIVITY CONCEPTS 7

properly in a productive environment. Thus, people are utilized where human skills yield the greatest payoff, and machines are likewise utilized where the payoff is also the largest. As suggested in Figure 1.1, the process exists in an environment containing factors for training, organization, leadership, and quality.

A boundary separates the internal parts of the productive system from its environment, and there should be a constant pressure to expand the boundary so that it is incorporated more fully into the total operational environment. A familiar example of this expansion is the accepted concept that quality emanates from the top through training and other management programs.

An increase in productivity achieved by combining the factors of

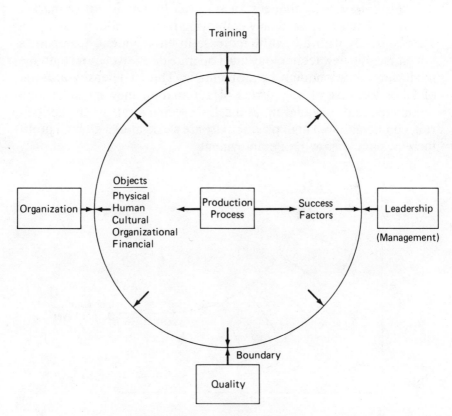

**Figure 1.1** A productive system

## 8 PRODUCTIVITY AND QUALITY

production in an advantageous manner is essentially the same as a shift in the production function, as suggested by Figure 1.2. This is the old "guns and butter" diagram that denotes the amount of each that can be purchased with a given amount of money. In cases such as this, the dislocation of the production function is normally associated with several major benefits:

- An increase in the standard of living
- Price stabilization
- Increased competitiveness
- Improved profitability

The end result, of course, is simply the fact that increased productivity leads to increased chances for survival. In a competitive market system, a progressive company employing new technology can operate offensively with a healthy increase in employment. Companies not employing new technology must operate defensively, and employment can correspondingly be problematic. Thus, a defensive attitude of labor towards various forms of automation may be in fact be unfounded and may end up as a self-defeating strategy. In the long run, companies with high productivity are associated with high profit increases and a growth in employment.

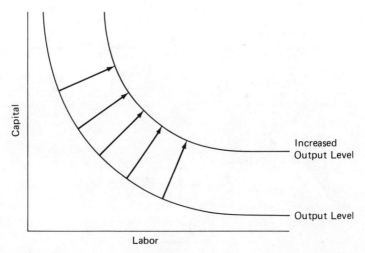

**Figure 1.2** Figurative increase in production, depicted as a production function, achieved through a productivity increase.

In some sectors of the economy, it is difficult to improve productivity in traditional ways. Service industries, such as some forms of government and human resources administration, are unfortunately typical examples, since most modern countries are in fact moving towards a more service-oriented society. Thus, the ability to work smarter rather than more efficiently can be a preferred approach.

## WORK

In one form or another, work has always been a primary ingredient in organized society. The concept of work is normally intertwined with the notion of a job as a means of obtaining goods and services that are deemed necessary for survival, recognizing of course that a luxury to one person may be a necessity to another. The concept of work expressed in this manner is not definitive since the type of work (i.e., physical versus intellectual) and its social overtones are not included. There are social reasons for working—such as satisfying emotional needs—that are important, and the question of ownership, working for others, and self-employment also makes a difference in the nature of the workplace and its effect on the individual. In the present context, a rather narrow viewpoint of the subject has been selected because of the emphasis on productivity, quality, and robots.

In the present context, *work* is regarded as those activities in which people engage that enables them to acquire needs and luxuries. The idea of exactly what a need is is clearly subject to debate, but the notion that work is performed for a reason other than recreation is intended in the present context.

It is necessary to look at two aspects of work that are important for improving quality and productivity:

- Differences in the perception of work by the worker and the manager
- Recognition that work has several components that must be considered in the context of an enterprise

In discussing work, the worker tends to emphasize inputs to the process in the form of time, thought, effort, and so forth. Management, on the other hand, tends to regard work in light of the level of output of the production process. Moreover, as work becomes less labor intensive, the management view tends to prevail.

The components of work are commonly regarded as:

- Effort expended in the work process
- Integrity of the work process
- Consistency of the work process

*Effort* refers to the physical and mental activities that constitute work. *Integrity* relates to the policies and procedures, together with effective resources, that add to the quality dimension of the work process. *Consistency* attributes yield a work situation that is reliable and dependable.

Whereas work, as a general concept, may properly be regarded in terms of inputs and outputs alone, *good work,* if such a thing exists, would be a reconciliation of worker and management views on the subject and a proper integration of effort, integrity, and consistency into the work process.

Today's professional managers can only start to solve the problem of instilling productivity and quality into the modern workplace. A total solution is essentially a matter of education, wherein young people can be taught to be creative and to strive for intellectual solutions to problems instead of the widespread emphasis in education on behavioral characteristics such as obedience and punctuality. In short, people have to be educated for "good work" at an early age.

## TECHNOLOGY AND EMPLOYMENT

Evolutionary trends in technology and productivity are commonly caused by mechanization and automation, in concert with new modalities related to energy, materials, and handling. Mechanization, of course, has aided the worker in the physical domain, and automation has aided the worker in the intellectual domain. In technological innovation, mechanization and automation are synergistically combined to achieve a given objective.

New technology impacts employment by reducing and creating jobs, which are effectively implemented through a complex system of employment dislocations. The number of blue-collar jobs is reduced by process control and automation, while at the same time, the number of jobs is increased through new products and support activities.

The number of white-collar jobs is reduced by office automation, data processing, and telecommunications, and is similarly increased by additional administrative positions and support personnel.

Changes in employment due to technological innovation can be conveniently placed in five categories:

- Changes to the workplace
  - Addition of new places
  - Adjustments to existing places
  - Elimination of existing places
- Changes to the plant
  - Addition of new plants
  - Expansion of existing plants
  - Replacement of existing plants
  - Shutdown of existing plants
  - Reduction in capacity of existing plants
- Changes to equipment
  - Introduction of new equipment
  - Replacement of existing equipment
- Changes to methods of production
  - Organizational change
  - Process changes
  - Informational changes (EDP, office automation, etc.)
- Changes in modality
  - New materials
  - New methods
  - New forms of energy

The inherent reasons for the above changes are as significant as the changes themselves, and not all of them are directly related to the "bottom line." Some of the most common reasons for the introduction of new technology are:

- Reduction of costs
- Changes in production
  - Introduction of new products
  - Elimination of older products
  - Adjustments to the amount of production
- Changes in the product
  - Quality
  - Characteristics

- Changes in the work force
  - Working conditions
  - Labor shortage
- Business conditions
  - Economic climate
  - Product demand
- Government regulation

Corresponding to each of the above modes of change and its associated reasons is an increase in the knowledge level of the participants. The processes of technological change are in themselves dependent upon greater degrees of specialization coupled with a blend of organizational awareness that was previously not needed. More than ever, the situation calls for a "strategy of change."

## ECONOMICS AND PRODUCTIVITY

An enterprise can approach the subject of productivity either offensively or defensively. The offensive team is deployed when it is necessary to increase profit margins or is possible to increase market share. The defensive team comes in when competitors have introduced new technology and it is necessary to guard one's position. The end result may be the same; however, the best defense may be a strong offense.

Profitability is directly related to productivity, and the reasons are obvious and straightforward:

- When the cost of labor and materials is the same among competitors, the company with the highest productivity can have the highest profit margins.
- Companies with the highest productivity tend to control the larger share of the market.

At the same time, it is important to recognize that productivity does not necessarily relate directly to cost reduction, but it does in fact imply the optimum use of available resources.

When productivity is interpreted to also include quality, then it affects the company in two ways:

- Effect on income
- Effect on cost

In the former case, higher quality can help to capture a greater share of the market and is complementary to input-output productivity. In the latter case, not doing things in accordance with requirements and specifications has a bottom-line effect called the "cost of quality." In general, quality includes both design and execution. The *cost of quality* is summarized as the cost of prevention, the cost of measurement, and the cost of failure. Clearly, operational procedures and methods of information handling can reflect poor quality as much as a manufactured product.

## PRODUCTIVITY, QUALITY, AND JOBS

On the surface, it would seem that productivity, quality, and jobs are three different things. In reality, however, quality is productivity. Defects are costly and ripple through an enterprise; they eat up profits like a swarm of locusts. Much of top management now regards quality as productivity, but this attitude has most certainly not filtered down to the shop floor.

The prevailing worker attitude is that pieces are one thing and quality is another, and moreover, "quality is a design problem anyway." This is partially true. Quality is a product design problem, but it is also a matter of job design, not only product or system design. If a job is too boring and results in poor quality, the remedy may be job redesign and not job simplification—as is commonly the case. Participative management is a catchall for a variety of management strategies, but its need has never been more in evidence than in times of concern over total productivity. The key issue is clearly a matter of integration of the diverse needs of production and quality control so that the goals of management and worker are mutually satisfied.

## BASIC INGREDIENTS IN A PRODUCTIVITY IMPROVEMENT PROGRAM

Heretofore, productivity has been regarded as a labor problem or as a case of increased mechanization or automation. It is only recently that informed management circles have involved the total organization in productivity and recognized the significance of the relationships among organizational units. Several key elements have been identified:

- *Top-management involvement* is needed in an effective productivity improvement program.

- *Worker involvement* is needed to identify areas where improvement can be made and possible courses of action for problem resolution can be decided upon.
- Quantitative aspects of *productivity improvement* through isolation and resolution should be enhanced by measurement and analysis.
- *Open communication* between management and workers should be encouraged, especially with regard to quality effectiveness procedures.
- *Productivity objectives* should be established and evaluated on a periodic basis.
- The total productivity program should be reviewed by disinterested parties to insure objectivity.

Many organizations have utilized the concept of a steering committee to govern the actions of managerial personnel and to insure that needed resources are available. Effective productivity programs require strategic planning involvement.

## SUMMARY

Productivity is in the news these days, but the context of consideration has changed in recent years. Total productivity is of prime importance, and automation and worker productivity, per se, must be placed in proper perspective. A balanced approach is required.

While the notion of productivity is rooted in history, the emphasis on the output function is a result of industrialization. In the present context, *productivity* is defined in real quantities as the relationship of the output of goods and services to the input of labor and capital. The concept applies to commerce and service industries as well as to manufacturing processes. Productivity measures are commonly applied to items that can be counted, such as units produced and cases resolved. Even when it is recognized that worker productivity is not total productivity, it is simply more convenient to concentrate on traditional items, such as worker satisfaction, job redesign, motivation and attitudes, and direct employee incentives.

Productivity is associated with division of labor and specialization. As such, this situation has both positive and negative aspects. In short, specialization is regarded as being desirable for the professional and dehumanizing and boring in the factory. Specialization has, in fact, resulted in a high level of individual worker skill and a relatively efficient production system. The competitiveness of modern industry is largely a result of effective production systems. The key result, of course, is to combine the factors of production in an optimum manner.

An important concept commonly incorporated into productivity consideration is the recognition that perceptions of work among participants tends to differ depending on a person's position in the organization. Proper attention to the fundamental aspects of work should result in integrity and consistency in the production process and in service industries and commerce.

In the long run, productivity gains and organizational success are directly related to the use of new technology— that is, if an organization will let it happen. Changes to the workplace are inevitable, but this phenomenon has occurred before in modern society.

There are economic benefits to productivity because it affects income and costs. However, in this context, due regard to the significance of quality as an objective, rather than a quantity to be measured, is necessary. The use of automation and some forms of participative management are important in achieving quality in products, services, and commerce.

Success in productivity involves the total organization. Key elements have been identified in a successful productivity program, and the need for strategic planning involvement has been recognized.

# 2 QUALITY CONCEPTS

## INTRODUCTION

Quality is important from a business point of view. Few would deny that fact. But there is more to the significance of quality than strictly business alone. Products and services constitute the very root of modern civilization. Society, as we currently know it, literally and absolutely depends upon the quality and the widespread availability of and access to essential products and services. Moreover, the need for manufactured products and services affects us physically, psychologically, and even intellectually. We can no longer return to a way of life wherein individuals, families, and communities are self-sufficient. Many products and services, such as the automobile and telephone that were once thought to be luxuries of a sort, are now commonly regarded as necessities. Clearly, the utility in both cases— that is, instances wherein luxuries become necessities—is directly related to quality.

Within the domain of quality itself, the manufacturing and service functions are only a small part of the total quality picture. Customer-stated needs and what the consumer will actually pay for must be determined by market research. Suitable products and services must be designed. Packaging, distribution, and sales must reflect the existence of quality. Service departments must sustain quality, and top management must create a climate for quality in administration, in operations, and in various forms of strategic and tactical planning.

There was a time in modern business when the responsibility for quality was thought to reside in the quality control department. Clearly, only in the most primitive of industries would this be the case today. Quality is a function of the total organization; in fact, we now refer to *total quality control* to emphasize the fact that quality begins with top management and permeates throughout the organization. Top management, however, must give more than lip service to a total quality control program. A committment to quality is actually needed, and the emphasis should not be diluted through various forms of compromise tactics whereby short-range economic analysis is used to lessen the impact of the total quality program.

This chapter delineates the scope of quality in today's enterprise.

## CONCEPT OF QUALITY

*Quality* is defined as the totality of attributes of a product or service that reflect on its innate capacity to satisfy a given set of needs. As a complete definition, this form may be lacking because quality means different things to different people. Some considerations that capture various aspects of the definitions are:

- Quality means "fitness for use."
- Quality reflects a customer's "acceptance criteria."
- Quality indicates the "grade of a product or service."
- Quality is "conformance to design objectives."
- Quality expresses "suitability for intended purpose."
- Quality signifies "reliability and maintainability."

Initially, one would think that it is easier to say what quality is not instead of what quality is; but even that leads to a dead end.

One relatively narrow meaning of quality is that it refers to the degree to which a specific product or service meets the expectations of a specific person. A custom-made suit, a portrait, a handmade pair of shoes, and a high-performance racing car are all examples of products in which quality reflects an implicit agreement between a manufacturer and an end user. Here, "fitness for use" for the intended purpose is the preferred meaning of quality. This definition, however, reflects the era of craftmanship in a preindustrialization period. By a stretching of the imagination, this definition can be forced to apply today—but that would be worse than fitting a square peg into a round hole.

The biggest problem in defining quality arises when an attempt is made to apply the word to different products, such as Mercedes-Benz and Chevrolet automobiles. Does a Mercedes have better quality than a Chevy? If you answer yes, then you are considering socially generated quality characteristics and reflecting a noninformed point of view. If you answer no, then you are properly regarding accepted determinants of quality, such as grade, conformance to design, suitability for intended purpose, and so forth. Clearly, it is better in this case to refuse to answer at all.

Certainly if a product does not serve its intended purpose, then it has poor quality, and this notion can be extended systematically to service organizations and other forms of commerce. In addition to basic service or serviceability, many persons prefer and also expect additional benefits that are commonly associated with quality. "Quality of service" is usually not anticipated from a bureaucratic clerk, but it is expected from your personal banker.

Because it sometimes becomes necessary to discuss quality as a technological concept, a more formal definition is required. In this and subsequent chapters, *quality* will be regarded as the degree to which a product or service serves its intended purpose, recognizing four key parameters:

- Quality of grade
- Quality of conformance to design objectives
- Quality of availability
- Quality of customer service

Each parameter reflects a particular aspect of the "intended purpose" and anticipates the dynamic nature of the quality function. Products and services are diverse in scope and in utilization. Some products are expendable and others are durable; services can be temporary or long lasting; some products deteriorate with age and others do not. Guarantees vary with regard to availability, reliability, and maintainability.

Because quality, as a concept, must survive in a climate of change, it has organizational implications at each stage in the life cycle of a product or service. These topics are fully amplified in subsequent sections.

## QUALITY LIFE CYCLE

Recognition of the fact that most products go through similar life cycles makes it relatively straightforward to identify areas where quality can be enhanced. Figure 2.1 gives the phases in the life cycle of a product. Each phase can contribute to quality, even though some phases are clearly more significant than others.

The *concept phase* involves determining the need for a product or

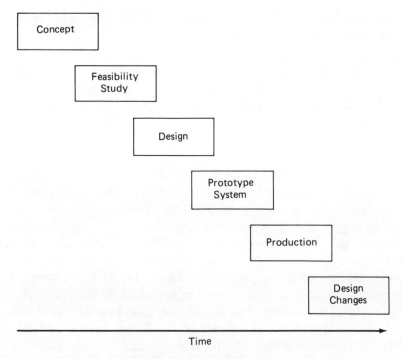

**Figure 2.1** Quality life cycle.

service. It is during this phase that serious consideration should be given to whether the product or service can be delivered with a given quality level. In fact, a specification of necessary quality attributes should probably be made initially so that this objective can be considered in the feasibility study for a new venture. The result of the concept phase is usually a report to the sponsoring department outlining a description of the product or service, a preliminary market analysis, and a statement of relevant business conditions.

The *feasibility study phase* effectively determines if the proposed product or service is in line with the goals of the organization. Inputs to the feasibility study include:

- Characteristics of the present product or service (if appropriate)
- Competitive analysis
- Viability considerations
- Organizational factors
- Financial considerations

## 20  PRODUCTIVITY AND QUALITY

During this phase, the functions performed by staff members normally involve the following tasks:

- Analyzing the current product or service
- Interviewing persons associated with the proposal
- Assessing impacts of the proposal
- Performing cost/benefit studies
- Preparing system proposal

Outputs of the feasibility study are a preliminary system description, an impact analysis, a cost analysis, and an implementation plan. The "cost of quality" should be incorporated into the feasibility study through its effects on income and cost.

The *design phase* involves the evaluation of alternative design proposals, the selection of a design strategy, the preparation of purchasing and materials requirements, and so forth. This phase is the cornerstone of an effective quality program. During the design phase, product or service objectives are evaluated and then reevaluated in light of the operational capability of the design entity. The design phase is necessary for instilling availability, reliability, maintainability, and overall consistency and integrity into the detailed product or service specification. Design principles, such as to decouple or integrate functional parts of a solution depending upon operational constraints, have existed for many years, but only recently has it been recognized that they possess a quality component as well.

The *prototype phase* is used to evaluate basic design factors, assess reliability, and assure integrity and consistency. It is during this phase that management can be alerted to potential quality problems through preproduction demonstrations and various forms of testing protocols. Design changes made at this point have a lesser impact than those made once full-scale production has begun.

The *production phase* is regarded as the "normal" production activity for customer delivery and service. The feedback cycle for quality assurance starts here, and employee participation plans, such as "zero defects" and "quality circles," show the most impact when initiated here. It is also in this phase that many forms of automation, such as industrial robots, are used. Clearly, automation, as a concept, achieves its greatest utility when products and services are purposefully "designed for automation."

The *design change phase* begins when product acceptance, consumerism, and customer relations activity result from products and services generated in the preceding phase. Traditional avenues of quality control—sometimes regarded as "inspection"—can influence the impact of production failures on product design. Clearly, the product quality that results from zero defects and poor design is markedly different from good design and poor quality control.

## QUALITY ASSURANCE

Quality assurance refers to the evaluation of quality activities with the objective of informing operational management of its current status. It is sometimes regarded as a quality audit, but many organizations now view quality assurance as an ongoing activity. It is through an effective quality assurance program that integrity and consistency of products and services are achieved.

Quality assurance is best implemented through a "quality assurance plan;" a plan is necessary because quality control activities can be subdivided to such an extent that quality soon becomes someone else's responsibility. A total *quality assurance plan* is a statement of organizational responsibilities for quality and requires input from the following sectors: marketing, design, materials (both procurement and control), manufacturing, and customer service. Moreover, quality assurance must be product or service oriented in order to be effective, as compared to plans that stress the behavioral aspects of quality.

The scope of quality assurance involves events that are both external and internal to the enterprise. External activities include the measurement and analysis of customer complaints and other field-oriented events. Internal activities include inspection reports, quality ratings, quality surveys and audits, and management reports regarding income and costs as they are affected by the presence or absence of quality.

One of the ongoing problems in quality assurance is the need to separate facts from opinion. Thus, the environment for quality assurance is of prime importance and has emphasized the need for participative management as a means of creating an atmosphere of quality awareness.

Specifically related to participative management in this context is the concept of a design review and a manufacturing plan. While the process of having one's tactical or functional plans analyzed by a peer group may be threatening to some engineers and managers, the result provides a "team approach" to quality that is difficult to achieve in other ways.

In manufacturing, failure analysis and nondestructive testing are useful methods of assuring quality and necessarily lead to a hierarchical set of organizational policies for the resolution of quality problems. The key benefit to be derived from a set of policies is that they apply as much to internal operations as they do to products and services.

## METHODS OF QUALITY CONTROL

Quality control is a process whereby the results of a process or service can be analyzed in order to make effective judgments concerning the characteristics of those results. While quality control normally involves acceptance, rejection, and other relevant criteria, it is a discipline in its own right involving measurement, analysis, and empirical judgment.

The measurement of quality is statistical or qualitative in nature and involves the following activities:

- Dissemination of knowledge concerning the *specification document*—that is, training, preparation of standards, development of acceptance procedures, and data collection methods
- *Inspection* of the product, explicitly or implicitly, through process control
- *Conformance* analysis
- *Acceptance* or *rejection*
- *Evaluation* of rejected products or services
- *Data recording*

The above activities necessarily include sampling, classification, and additional forms of analysis as they relate to storage, transportation, and customer acceptance. The return, reject, or complaint ratios in relation to price and time are also significant.

Statistical methods in quality control have a long and successful history. Control charts, basic statistical analysis, experimental design, hypothesis testing, correlation and regression analysis, and other forms of forecasting and prediction are widely known and commonly

QUALITY CONCEPTS 23

used. The use of qualitative methods is less widespread because of the propensity in modern industry for collecting numerical information and then processing it.

Two qualitative methods are particularly useful in quality control:

- Cause-and-effect diagrams
- Pareto diagrams

A *cause-and-effect diagram* acknowledges the recognition of an effect and serves to delineate the causes, as suggested by Figure 2.2. Major causes—factors—are identified first and are subsequently broken down into subcauses. Once factors and their relationship to other factors are exhaustibly identified, then they can be analyzed to determine which ones contribute most significantly to the observed effect. A *Pareto diagram* (see Figure 2.3) is a means of arranging quality factor and quality cost data according to priority or quality cost

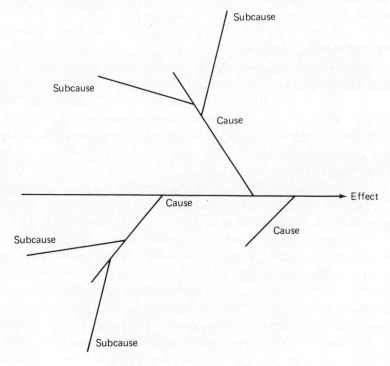

**Figure 2.2** Cause-and-effect diagram.

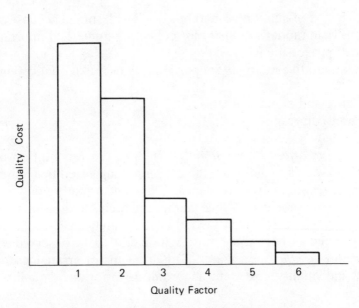

**Figure 2.3** Pareto diagram.

value—as the case may be—so that factors that contribute most heavily to the cost of quality can be handled on a priority basis. Both qualitative methods are covered in more detail in later chapters.

A topic that frequently arises in quality control is the significance of and the difference between accuracy and precision. This subject comes up when an instrument or process for measurement does not give a true reading. *Accuracy* is a measure of the extent that the *average* result of a process or service deviates from a true, a desired, or an expected result. An instrument may or may not accurately record a true value of a physical phenomenon. A service rendered may or may not produce the desired outcome. Accuracy refers to the fact that, on the average, whether values are high or low, big or small, etc., what counts is the average value and the deviation of values from the average. Accuracy reflects deviations caused by calibration deficiencies in physical processes and misalignment of interpretation in a service domain. *Precision* refers to the repeatability of the measurement or interpretation techniques when applied to true, desired, or expected results. Thus, a probability distribution curve (as suggested in Figure 2.4) with a low deviation reflects high precision, and a distribution with a high deviation denotes low precision. Similarly,

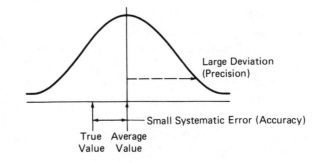

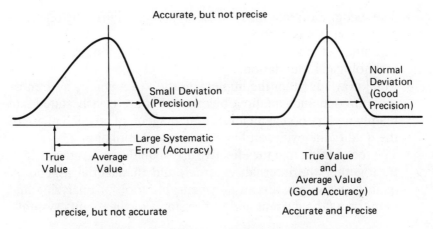

**Figure 2.4** Conceptual difference between accuracy and precision.

the difference between a true value and the average value is a reflection of accuracy.

Methods of quality control will have a relatively small impact on total quality control unless questions of accuracy and precision are adequately resolved.

## QUALITY AWARENESS

Whereas quality assurance is concerned with planning and maintaining product quality, quality awareness involves the ongoing task of insuring that an effective and continuous channel of communication exists to inform all persons involved in the quality life cycle concerning quality-related events. The stimulus for quality awareness comes from both inside and outside the organization. A total quality control program is clearly the starting place for quality awareness from within the organization. Consumer groups and the government have created a general concern for quality, and the pressures of inflation have emphasized the importance of quality for the consumer from outside the organization. The current state of affairs is reflected in the following exclamation: "If I have to pay that much for a new car, it had better be perfect!" Thus overall expectations are higher now than they have been in the past.

The basis for quality awareness stems from four key concepts:

- The recognition that the control of quality is currently regarded as everyone's business
- The understanding that quality achievement requires a strong technological foundation
- The knowledge that the implementation of a quality awareness program demands quality objectives that are clearly stated and readily understood at all levels of management, at each phase of the quality life cycle, and in every organizational unit
- The recognition that the effective use of quality indicators, which tie together the internal, external, and managerial aspects of quality awareness, are an indispensable tool for analyzing and monitoring the current and future states of the quality system

The fact that total quality is a complicated combination of design, process, and integrity considerations is fundamental to the four concepts. The emphasis in the last category is, of course, on people.

The key to quality awareness is the creation or development of an operational environment that has a strong enough foundation to support quality programs incorporating a variety of methods and concepts.

Probably the most significant building block of quality awareness is *participative management,* defined informally as the involvement of people in the creation, development, and quality of a final product. The basic idea is obvious; people are simply more committed to a product in which they are involved. As a concept, participative management contributes to quality awareness as a communication technique and also as an operational modality.

Through participative management, the status of the quality situation is constantly being communicated through small-group activities, and involvement is effectively being solicited in two primary avenues:

- Establishing quality improvement goals
- Recognizing methods of achieving quality goals

Thus, the very process of participative management is, in itself, a means for achieving quality awareness.

Almost any quality program can result in enhanced quality awareness. Some of the better-known quality programs are "zero defects," "awareness teams," and "quality circles"—to name only a few. The technique stressed here is the concept of quality circles, which has applicability to a wide range of processes and is not restricted to manufacturing operations. As implied in the section entitled, "Methods of Quality Control," both quantitative and qualitative information is commonly available on quality and the quality circle approach can effectively utilize both forms. Regardless of the quality program selected or the characteristics of the data on quality, the subject of quality indicators cannot be ignored. Indicators operate in a fashion similar to economic and medical indicators; they provide information on the state that a system is in. Collectively, the formal recognition of a set of quality indicators is an efficient means of initiating "total" quality awareness.

## QUALITY POLICY, OBJECTIVES, AND PLANNING

The difference between an effective quality program and a marginal operation is the result of three related activities:

- Formulating policy
- Establishing well-defined objectives
- Quality planning

Clearly, these activities are directly analogous to the manner in which things normally get done in most organizations; they serve to establish broad guidelines for action and impact practically every dimension of the management process. Because policies, objectives, and effective planning are needed at all levels in an enterprise, it is important to establish avenues in which a quality program differs from other business areas and from the organization as a whole.

On the surface, establishing a quality policy may appear to be obvious, but market competition and the dynamics of consumer behavior serve to complicate the situation. Basically, a quality program can serve to either improve quality by promoting change in the present operational environment or to maintain an existing quality level by preventing change in the operational environment. The former case may evolve to the latter case, when satisfactory results are obtained, and serves to enforce the underlying knowledge that quality is as dynamic an endeavor as is business itself. Nevertheless, a well-defined set of quality objectives can serve as input to the policy formulation process. The list of possibilities in this regard is probably open-ended. Three obvious cases surface. When one is dealing with standardized products where performance levels are determined by established convention or by law, then a primary quality objective is to sustain required levels of quality and face competition through technical services, price, or product availability. When one is providing customized products, as in the chemical process industry, quality is effectively determined by the customer specification and an organization must be competitive through technical service or processing efficiency. Clearly, quality concepts apply to technical services and to internal processing operations as they do to manufactured products. The final case concerns the dominant competitive environment wherein price and quality determine a company's market position. The resulting objective to be a quality leader, a "functional" leader, or an innovative leader governs the quality policy process, which is designed in most cases to achieve change in the direction of higher quality.

In formulating a quality policy, the policymaker must be concerned with various business practices:

- The age-old belief that the key objective of production is to keep the machines busy
- The desire to be competitive, regardless of quality

Along with each of these practices is the underlying theory that the function of the sales department is to sell a product, regardless of quality. In order to counter traditional issues such as these in quality control, it is usually recommended that clearly defined quality objectives be materialized in the form of written policy that clearly presents an organized approach to quality through an authoritative style and a uniform perspective. Written policy has an innate power that stimulates forethought and planning and consolidates management's thinking on the subject. Thus, an effective quality program, supported by written policy objectives, can break down negative attitudes and steer the organization in the direction of overall consensus on quality issues.

In planning to meet quality objectives, objectives are broken down into constituent elements that include

- Control (i.e., steering)
- Diagrams
- Standards
- Organization

Although the need for quality in the modern enterprise is universally recognized, the interdepartmental nature of its activities and the effective utilization of technology tend to obscure an obvious fact: that planning for quality matters is as significant—from the standpoint of management action—as the quality process itself. In many cases, the practice of quality achievement is initiated through a carefully selected pilot project that serves to "prove" the various concepts.

## QUALITY ORGANIZATION

The responsibility for quality organization lies with executive management, which is also responsible in the management hierarchy for organization in general and quality in particular. Quality organization has evolved through three major steps:

- As the responsibility of the production foreman himself or delegated to one of his workers
- As separate quality inspection limits organized under a chief inspector
- As a full-blown quality department, at the same level as marketing and manufacturing, and incorporating the functions of inspection, quality control engineering, reliability, and quality assurance.

While quality organization is as different among enterprises as any unit could be, the final evolutionary step (given above) provides a modern approach whereby each contributing factor to quality can be organized individually to satisfy local needs. Thus, there can be organization for quality assurance, organization for reliability, organization for inspection, and so forth. Flexibility and coordination are inherent in the fact that each quality unit in the total quality organization reports to a common manager.

As by-products of the modern approach to quality organization, data reduction can be handled as a service (or staff) function. Special committees for doing coordination need not explicitly exist, because the structure of the organizational units provides a common thread of interaction.

Many quality people feel that the key to quality organization is not the structure of the organization but rather the practice of participative management that filters down through the entire organization. The most popular form of participative management is the application of quality circles, covered in Chapter 3.

## SUMMARY

Quality is important from a business point of view because products and services constitute the very root of modern civilization and society literally depends upon the quality and widespread availability of and access to them. Many products and services that were once considered luxuries are now regarded as necessities, and a loss of them would affect us physically, psychologically, and intellectually. Clearly, the utility of most products is associated with their quality. Although quality is the backbone of an effective business plan, it can only be achieved through total quality control that involves most organizational units, including marketing distribution, service, and finance—in addition to manufacturing and engineering.

Quality is defined as the totality of attributes of a product or service that reflect on its innate capacity to satisfy a given set of needs. Several associated considerations to this definition are

- Fitness for use
- Acceptance criteria
- Grade of product or service
- Conformance to design objectives
- Suitability for intended purpose
- Reliability and maintainability

In some respects, quality is "all of the above."

Most products go through a similar *life cycle* that makes it possible to identify areas where quality can be enhanced. The various phases are summarized as:

- The concept phase
- The feasibility study phase
- The design phase
- The prototype system
- The production phase
- The design change phase

Each phase contributes in a unique way to the total quality of a product or service.

Quality assurance refers to the evaluation of quality activities with the objective of informing operational management of its current status. Compared to a quality audit, quality assurance is an ongoing activity. Quality assurance is best implemented through a "quality assurance plan," incorporating events that are both external and internal to the enterprise. Quality awareness through various programs of participative management enhances the scope of quality assurance and provides a team approach to the subject.

The methods of quality control include both quantitative and qualitative techniques that involve the following activities and events:

- The specification document
- Inspection
- Conformance analysis
- Evaluation
- Data recording

Statistical methods in quality control are well developed and have a long and successful history of providing support for quality control. Two recently recognized qualitative methods are

- Cause-and-effect diagrams
- Pareto diagrams

In order to be totally effective, methods of quality control must incoporate measures of accuracy and precision.

Quality awareness involves the ongoing task of insuring that an effective and continuous channel of communication exists to inform all persons involved in the quality life cycle concerning quality-related events. The impetus for quality awareness comes from within the organization based on a general concern for quality established by consumer groups and the government. A key element of quality awareness is participative management that provides a channel for communication through small-group activities. Two primary areas of involvement are:

- Establishing quality improvement goals
- Recognizing methods of achieving quality goals

Although almost any quality program can result in improved quality awareness, three of the better known programs are zero defects, awareness teams, and quality circles.

Policy formulation, the establishment of well-defined objectives, and quality planning spell the difference between a marginal and an effective quality program. Quality policies serve to either improve quality by promoting change in the present operational environment or to maintain an existing quality level by preventing change. Clearly, the quality picture is as dynamic as business itself. Some of the factors that influence the delineation of quality objectives are standardization, customized products, and a company's competitive market position. In order to achieve effective planning, objectives are commonly broken down into the following elements:

- Control
- Diagnosis
- Standards
- Organization

From the vantage point of upper management, the planning for quality matters can be as significant as the quality process itself.

Quality organization has evolved through three major stages:

- Foreman centered
- Inspection centered
- Department oriented

In the last case, full recognition is given to the fact that the modern view of quality incorporates inspection, engineering, reliability, and quality assurance. Flexibility and coordination are inherent in effective quality organization.

# PART TWO
# QUALITY CIRCLES

# 3 QUALITY CIRCLE PRINCIPLES

**INTRODUCTION**

A *quality circle* is a small group of workers who meet regularly on a voluntary basis to analyze problems and recommend solutions to management. The organizational sphere of activity of quality circle participants is the area in which they work. The concept of a quality circle originated in Japan, to assist in solving quality problems in manufacturing, and has since been applied to a variety of related managerial problems throughout the world in diverse disciplines. In fact, quality circles were originally known as "Quality Control (QC) Circles" in Japan and are still sometimes called QC circles there. Clearly, the primary application domain of quality circles is the area of quality achievement; however, quality achievement can take a variety of forms, ranging from worker behavior to management cooperation. Moreover, quality achievement may or may not involve high technology and is as applicable to banking as it is to manufacturing. Quality circles have been used in both private and public enterprises in countless variations. To a greater or lesser degree, most organizations adopting the quality circle concept have adapted it to their particular needs.

The basic philosophy underlying quality circles is that quality awareness through participative management can not only identify problem situations but can also assist management in solving them. Clearly, people know their jobs best and want to contribute to the success of their company, if only given the opportunity to do so. With regard to Theory X and Theory Y management styles, which are well known in their own right, the use of quality circles is a Theory Y approach. A quality circle is a form of group dynamics; members collaborate to identify and solve problems, and the group members have a synergistic effect on each other's work behavior and additionally to their individual contributions to collaborative quality circle activities.

Most descriptions of quality circles begin with a heavy dose of history, with the commendable objective of attempting to describe the salient benefits of quality circle activities and preserve the origins of the techniques. Although there is certainly some benefit in history, it is also important to recognize that cultural differences between countries imply that an approach that works in one country may not be directly transferable to another country, lessening the overall benefit of a historical perspective. As an alternative and perhaps a more direct approach to quality circle dynamics, the following list of quality circle attributes is presented:

- Participation in quality circle activities is voluntary.
- A quality circle is a small group, composed of 4 to 6 people in small shops, 6 to 10 people in medium-sized shops, and perhaps 8 to 12 people in larger shops.
- Quality circle members do similar work or their work is related in a logical sense so that members normally work as part of a team to achieve a common goal.
- Quality circles meet regularly to discuss and solve problems that they identify or that are proposed to their circle leader.
- Every quality circle has a formal leader responsible for the operation of a circle. The leader is usually a supervisor, for reasons given below, and is given special training on quality circle activities. Alternately, the quality circle leader is called a "circle coordinator," although this title is sometimes used for other purposes.
- A quality circle program has a "quality circle facilitator" who is responsible for monitoring, guiding, promoting, coordinating, training, and communicating the essentials of quality circle techniques. The facilitator provides an interface between the quality circles and other organizational groups, including a management steering committee.
- A management steering committee establishes objectives, policies, and guidelines for quality circle activities and supports the quality circle system through adequate resources and management awareness.
- An optional policy committee oversees the operation of multiple steering committees and determines the overall level of management commitment to the quality circle program.

Thus, a quality circle program is a well-defined set of policies, procedures, and people established to increase the effectiveness of the total work environment. The key element in a quality circle program is participation, and the basis for participation is knowledge and training. Every participant in a quality circle program—ranging from the circle member to the policy committee director—receives training or familiarization appropriate to his or her degree of involvement in the system.

## ORGANIZATION FOR QUALITY CIRCLES

No change to the organizational structure is needed to achieve a successful quality circle program. The policy and steering committees represent traditional management functions. Similarly, circle members and leaders assume their normal roles in the organization—supplemented, of course, by quality circle activities. The only new position is the quality circle facilitator, who oversees the operation of several circles. In some organizations, the facilitator is a member of the human resources department—perhaps the Management Development/Organizational Development (MD/OD) staff. In others, the facilitator can be a member of the Education/Training, Operations, or Quality department. In any case, the result is the same. The facilitator works through the existing organization to establish a quality circle program and subsequently makes it possible for the results of quality circle activities to be recognized and integrated into the operational basis of the organization.

Figure 3.1 gives an organization chart for quality circle activities. At a glance, it is evident that member involvement is crucial to an effective program. Every link in the chain, however—including the development of suitable policies and procedures—is needed for the results of quality circle activities to show benefits for the organization. In fact, an avenue for the implementation and integration of the results of circle activities must exist through complementary policies and procedures in order for a quality circle program to survive as an organizational resource. Without effective support, a quality circle program can easily be regarded by circle members as another management attempt to increase productivity without offering tangible resources. Clearly, quality circles should increase productivity by "working smarter," but initially the key factor should be participation.

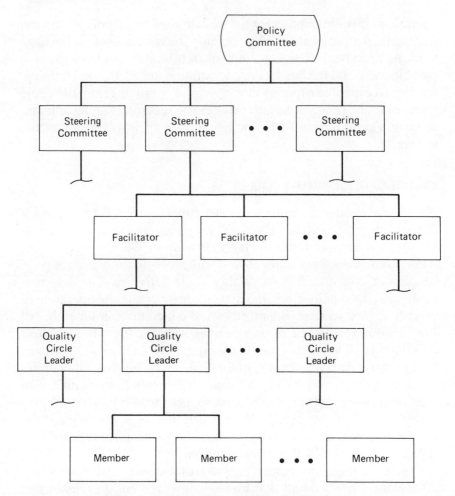

**Figure 3.1** Organization of quality circle activities.

Figure 3.2 gives a list of conceptual job descriptions in a logical quality circle chain. An optional *coordinator* position—not to be confused with the quality circle leader mentioned earlier—is included as a means of materializing the directives of the steering committee in some organizational settings. In most cases, the directives can easily be handled by the facilitators.

In addition to the primary sources given previously, the facilitator and coordinator positions can readily be handled by staff assistants in technical areas, such as banking, data processing, and manufacturing.

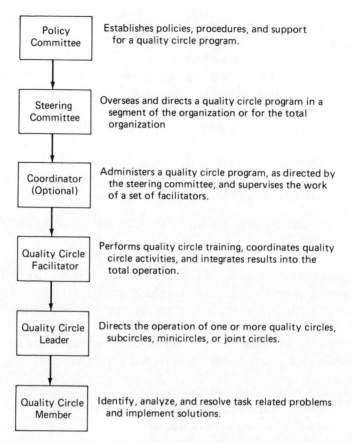

**Figure 3.2** List of conceptual job descriptions in a logical quality circle chain.

Because a quality circle works from within the organization, any member of the existing organizational structure (i.e., the department) has a distinct advantage over external personnel in establishing an effective program.

In the following sections, a role model is given for each of the participants covered here.

## POLICY FORMULATION

The basis for quality circle policy formulation is the specification of purposes and, eventually, objectives for the program. The purposes of a quality circle program are threefold:

- To enhance the capability of an enterprise to achieve its intended purposes—be they profit, service, or any other form of achievement.
- To provide motivational factors for human fulfillment and permit the process of participating in work to be a meaningful human experience. In short, this is the process of regarding people as humans and machines as objects and allowing the two modalities to be integrated only when necessary for achieving an intended purpose.
- To fully utilize and expand the capabilities of individuals through training and participation, and by providing a basis for using the innate creativity of people.

Clearly, the three purposes are related, as are most organizational activities.

Based on these fundamental purposes, a set of objectives can be developed that serve as input to the processes of policy formulation. Objectives are necessary for establishing the operational limits of a policy set. Although many major and even minor objectives can be listed, it is particularly important to present only those that are reasonably independent of management style, organizational structure, and the type of enterprise. Five universal objectives are given:

- To improve productivity and quality
- To improve communications
- To enhance cooperative behavior
- To provide an efficient operation
- To enhance morale

*Productivity and quality* can be achieved through a strong workshop that encourages problem detection, analysis, and resolution. Looking at the total picture, workers best know where their productivity can be increased and recognize operations that contribute to lower quality or even poor quality. Clearly, a quality circle program can contribute to high productivity and quality when integrated into the workshop routine.

Effective *communication* is necessary for the foreman, supervisor, or manager to achieve a positive influence and a state of control over the employees. It is obviously necessary for a workshop or office to perform in a given way—regardless of the enterprise. In a controlled workshop, potential problems are nipped in the bud by observance of established and accepted standards. When appropriate corrective measures are taken to abnormal conditions, causes that would ordinarily result in out-of-control situations are resolved before disastrous effects occur. Control can be achieved through automation—such as the use of industrial robotics—and also by the establishment of realistic standards and operating procedures.

People cooperate when they understand the job, the requirements and standards, and each other. When people engage in group problem solving, as with quality circles, they enhance *cooperative behavior.* Because quality circle activities are voluntary, people cooperate because they want to contribute to their own success. One of management's greatest failings in the post-Theory Y era is the thinking that workers may in fact be loyal and hardworking but nevertheless are clearly subordinate in their ability to make an intellectual contribution in their ability to make an intellectual contribution in the area of decision making. Nothing could be further from the truth, since workers know their individual jobs and the associated operational environment better than anyone. Moreover, attitudes of this sort, which also exist between levels of managing, have a real tendency to become self-fulfilling prophecies. What has been lacking, obviously, is the mechanism for allowing people to contribute, and the use of an effective quality circle program now appears to be the answer.

A key objective of quality circles is to assist in providing an *efficient operation.* This objective can be regarded as the same as "productivity and quality," but it can also represent a way of looking at the costs inherent in preventing, appraising, reducing, and resolving failure. Failure costs, regardless of whether they reflect service or product failure, can be determined through an effective reporting system. Needed information is obtained by monitoring, analysis, summarization, and communication. Monitoring, analysis, and summarization rely on quality circle techniques, and communication is a group process. Through participation, the formulation of the quality measures helps to insure the success of a program to reduce failures and improve efficiency.

Looking at the enterprise in a collective sense, composed of management and workers, it has been recognized in the past that quality circles form (or are formed) for three reasons:

- Training
- Problem solving and generating ideas
- The natural tendency of people with a common interest to band together

It has also been recognized that the overall contribution by workers and foremen to the solution to quality control problems does not extend to more than 10 to 15 percent of the problems. Then, why all the fuss over the simple concept of "getting together to solve problems?" The answer is improved *morale* and an increased desire to implement solutions to recognized problems. Potential benefits are evident from the two main ways that quality circles are formed:

- As employees from the same work unit, but not necessarily performing the same task
- As employees performing the same task, such as word processing, in different organizational units

In either case, employees can share work-related problems, resulting in a higher level of effectiveness. Quality circles do not result from high morale, but rather, high morale results from quality circles. Although quality circles should not be regarded as a management tool for controlling the workplace, a higher level of management control is pragmatically achieved through worker participation.

The set of purposes and then objectives feed into a strong policy statement that can give the quality circle program direction and continuity. This approach provides a guide to management activity with regard to:

- Consistency of management action
- Delegation of responsibility
- Uniformity of follow-up

Overall, a good policy structure eliminates confusion and reduces misunderstanding. Because of the diversity of tasks that may be involved and the number of people, the policy set should be reasonably short and well defined. Table 3.1 gives a sample policy set that can serve as a starting place for any organization initiating a quality circle program.

### Table 3.1  Sample Policy Set for Quality Circles

Employee participation:
 1. Employees are free to join in, not join in, discontinue, or continue participation in a relevant quality circle.
 2. An employee may suggest problems for solution to a relevant quality circle or to management for general solution.

Management support:
 3. Management will encourage quality circle membership and participation.
 4. Management will provide resources for quality circle activities.
 5. Management will publicize quality circle activities.

Management participation:
 6. Management will respond to quality circle requests for information, recommendations, and management participation.
 7. Management will implement approved quality circle solutions.
 8. Management will provide training support.

Quality circle activities:
 9. Quality circle members will follow techniques and procedures established in the quality circle manual.
 10. Quality circle members will recognize the anonymity of members.
 11. Quality circle members will respect an employee's voluntary right of participation.
 12. Quality circle members collectively have the right to accept or reject problems submitted by management, other circles, or other members.
 13. Quality circle members are responsible for identifying, analyzing, and implementing problem solutions.
 14. Quality circles may submit problems to management for acceptance.
 15. Quality circle members will collaborate on work-related problems.
 16. Quality circle members will respect the integrity of other circle members.
 17. Quality circles will present periodic reviews of activities to management.

Organization:
 18. Quality circle organization will consist of an optional policy committee, one or more steering committees, optional coordinators, facilitator, quality circle leaders, and quality circle members.

Restrictions:
 19. Quality circles will not address subjects related to employee complaints, personnel policy, compensation, labor relations, and personal employee problems or characteristics.
 20. Management will not penalize or impose restrictions on employees as a result of participation in quality circle activities.

## IMPLEMENTATION

Implementation of quality circles is straightforward and practically without risk. In fact, the simplicity of the approach to participative management can easily belie the potential benefits that can be derived. Two similar implementation plans are described in the literature: one by Ingle (1982) and the other by Beardsley (1977). Both can be properly regarded as top-down approaches, since an implicit assumption is made that the overall concept has already been sold to top management. The two plans are summarized, in spite of their similarity, in Tables 3.2 and 3.3.

An alternate scenario for implementation, of course, is to use a bottom-up approach. A department manager or foreman hears about quality circles and decides to try the idea in a problem area. In some cases, members of the human resources staff can even recommend the concept to managers with quality or personnel problems. As a result, a pilot project is started—preferably with top management's blessing. Then, if good results are obtained, the concept can be dispersed within the organization through the espousal of a sponsor. Two key advantages of the bottom-up approach is that start-up costs are lower than with top-down and it is less likely to give the impression that it is another management strategy to increase worker output.

## QUALITY CIRCLE PARTICIPANTS

Four classes of employees participate directly in quality circle operations:

- Steering committee
- Facilitator
- Circle leader
- Circle members

It is important to note, however, that as the results of quality circle activities affect the organization, management and nonmember employees are also affected. Clearly, when the results of quality circle activities are accepted and implemented by management, the organization must change. However, the very presence of ongoing quality circle activities also changes the operational dimension of the organization through "people building", creativity, team spirit, and the training inherent in the processes.

### Table 3.2  The Ingle Approach to Top-Down Implementation

| Step | Description |
|---|---|
| 1 | *Select a two-person team—one from quality control and one from industrial relations.* (The team will do the necessary research and establish the basis for the use of quality circles in the organization.) |
| 2 | *Research.* (The team will collect and summarize relevant literature on the subject.) |
| 3 | *Attend a seminar.* (The team, as well as other interested parties, should attend an outside seminar, given by experts, on quality circle methods and case studies.) |
| 4 | *Observe quality circles.* (Visit an organization that uses quality circles and observe the group dynamics of a circle in operation.) |
| 5 | *Decide to start.* (Results are summarized and presented to top management for final approval.) |
| 6 | *Select a facilitator.* (This is a key position involved with planning, coordinating, training, and follow through.) |
| 7 | *Form a steering committee.* (The steering committee provides overall direction for the program and substantiates resource requirements.) |
| 8 | *Develop plans and goals.* (Establish objectives and goals and associated strategic and tactical plans.) |
| 9 | *Present the plan to management and the union.* (Communicate the philosophy of quality circle participation and the detailed strategic and tactical plans to all levels of management and pertinent union officials.) |
| 10 | *Develop training material.* (The material should include presentations, manuals, techniques—a complete scenario for quality circle activities.) |
| 11 | *Present the concept to the group.* (The quality circle groups involved in implementation should be consulted and familiarized with the objectives and goals.) |
| 12 | *Start training.* (The training plan should be prepared and actual training of members and leaders should be initiated.) |
| 13 | *Form circles.* (Working quality circles should be formed in the latter stages of training.) |
| 14 | *Review monthly progress.* (A status report covering all quality circle projects should be prepared on a periodic basis.) |

### Table 3.3 The Beardsley Approach to Top-Down Implementation

| Step | Description |
| --- | --- |
| 1 | *Discover.* (A person in the organization experiences a need and relates it to the quality circle concept. At this time, the ideas are reasonably well known.) |
| 2 | *Research.* (Collect reference material.) |
| 3 | *Attend seminar.* (Attend a seminar given by a specialist on the subject. This step signifies a transition from the research to the implementation stage.) |
| 4 | *Witness quality circles in action.* (Visit an organization using quality circles to discuss the concept and observe a circle in operation.) |
| 5 | *Decide to start.* (Discuss the quality circle with concerned employees and summarize results for top management's approval.) |
| 6 | *Consult.* (Decide whether an outside consultant is needed to assist in planning, training, and analysis.) |
| 7 | *Select a facilitator.* (This is the person responsible for coordinating and planning the quality circle program.) |
| 8 | *Perform base-line measurement.* (The "present position" should be recorded in terms of cost, productivity, quality, tardyism, absenteeism, and so forth.) |
| 9 | *Create a steering committee.* (A high-level group of staff and management personnel, responsible for supporting the program and giving it direction, should be formed.) |
| 10 | *Develop implementation plan.* (The facilitator, possibly in conjunction with a consultant, develops the general plan for introducing quality circles to the organization. The plan should be coordinated with concerned management and approved by the steering committee.) |
| 11 | *Present implementation plan to management.* (The plan should be presented to all levels of management for agreement and familiarization. This step should be supported by the steering committee.) |
| 12 | *Develop training material.* (This step involves the establishing of training objectives, materials, and a realistic plan for acquainting circle leaders and members with quality circle concepts and techniques.) |
| 13 | *Conduct leader training class.* (Quality circle leaders are introduced by the facilitator and consultant, if appropriate, to quality circle concepts, methods, and member training. This is an ongoing activity.) |
| 14 | *Start the quality circle.* (This step is the preparation of quality circle activities in designated areas, and includes scheduling, personnel selection, planning, and notification. |

*(continued)*

**Table 3.3** *(Continued)*

| Step | Description |
|---|---|
| 15 | *Introduce to potential members.* (This is the "kick-off session" to acquaint circle members with the reasons and objectives in forming a quality circle. This meeting is held by the quality circle leader with the assistance of the facilitator and a key management person.) |
| 16 | *Conduct member training.* (Circle members are trained in quality circle activities in three or four sessions by the facilitator and the leader. In the fourth or fifth meeting, training is tapered off and the group begins to identify, analyze, and solve practical problems.) |
| 17 | *Review program.* (Quality circle activities should be reviewed periodically by the steering committee, top management, and the policy group.) |

## Steering Committee

From an organizational perspective, a quality circle program cannot succeed without the support of a steering committee. Local solutions to particular management problems can be developed through participative management, but their overall impact will be less than if a total quality circle program is adopted. On the other hand, an overzealous domination by the steering committee of quality circle activities can easily kill an otherwise successful program. So, management must walk a tightrope between the two extremes, with frequent adjustments to changing conditions. The greatest danger has been mentioned previously, and it is the possibility that prospective circle members will jump to the conclusion that quality circles are just another management "trick" and decline participation. Therefore, the committee must give careful attention to the policies that will guide circle participation and operation.

The steering committee should include staff personnel as well as managers from the broad spectrum of organizational functions. Rather than "make it happen," the steering committee's logo should be "let it happen." Table 3.4 gives a summarization of quality circle functions. Other than administrative tasks, the committee should focus on objectives, areas of concentration, personnel assignments, and operation of the program overall. Union participation should be encouraged but not expected. Some unions prefer to observe rather than participate. It is important that quality circle activities not overlap

**Table 3.4  Steering Committee Functions**

- Develops policies and guidelines
- Establishes a regular monthly meeting time
- Identifies to whom the steering committee will report
- Prepares objectives and prepares implementation plan to achieve these objectives
- Provides guidance and direction
- Determines funding arrangements
- Determines areas on which circles can focus attention
- Promulgates quality circles throughout the organization
- Identifies milestones in circle progress
- Determines or reviews start dates of circles
- Identifies and approves leaders for circles
- Establishes qualifications for facilitators
- Decides on frequency of circle meetings
- Decides how divisions and departments will learn about quality circles
- Schedules familiarization presentations and orientations
- Selects facilitators
- Establishes how circles' efforts will be recognized and rewarded (if appropriate)
- Periodically reviews program milestones
- Attends management presentations (of quality circle activity)
- Meets periodically with facilitators

with union jurisdiction, and a policy to this effect is normally established.

The size of the steering committee is not as important as the fact that it represents the total organization and meets regularly. Specialists indicate that a minimum membership of 5 and a maximum of 15 is ideal.

In small organizations (or small quality circle programs), the facilitator should be a member of the steering committee. For large programs, this is clearly impossible because several facilitators would normally be needed.

Training of the steering committee is minimal but important. Policies and day-to-day operations are largely dependent upon a particular organization, so a briefing on quality circle concepts is all that is normally needed. A consultant, the sponsor, or an experienced facilitator can easily perform the briefing. It follows that familiarization should be informal, in line with the key objective of participative management.

The chairman of the steering committee can be appointed by top management, be elected, or be a member of top management itself. In any case, the person should be democratic, open to suggestion, and definitely not a "table pounder." Nothing can kill a quality circle program more quickly than a dominating steering committee, and a domineering leader can severely influence the committee in that direction. In his or her own way, the chairman of the steering committee has to be a salesperson to counteract the negativism normally associated with something new. Two aspects of selling are required: selling on the merits of the program and selling to obtain involvement. Many managers do not want to be guinea pigs and feel comfortable with historical methods—even though they may be largely ineffective.

There are also some management decisions associated with the steering committee's activities. Most decisions relate to the facilitator and involve selection and full-time versus part-time participation. Policies for selecting leaders and establishing recognition and rewards are also significant.

**Facilitator**

The facilitator is the key person in a results-oriented quality circle program, by promoting the quality circle concept and coordinating the activities of leaders, the steering committee, and management. The facilitator coordinates within circles and between circles and serves as the interface to other departments—such as engineering or marketing. When assistance is needed from outside groups, the facilitator arranges it. The facilitator reports to management and the steering committee on the status of the quality circle program in the organization. Facilitator functions are summarized in Table 3.5.

The facilitator serves also as a trainer for quality circle leaders and possibly the circle members. In essence, the facilitator is the quality circle expertise, and this expertise is manifested through the preparation of training materials and promotional literature.

The facilitator effectively sets quality circle standards for the organization and is knowledgeable of analysis techniques and methods of presentation.

Personal qualifications of a successful facilitator vary widely depending upon the parent organization. Overall, however, the facilitator is a people-oriented person, challenged by the process of working through people to get the job done. While the facilitator need not be a manager, the pay scale for the position should be roughly equivalent to a technical employee.

There are two basic prerequisites for the facilitator position: coordination experience and a background in the primary function of the organization—such as manufacturing or banking. The facilitator should

#### Table 3.5  Facilitator Functions

- Trains leaders
- Promotes the quality circle program to employees, supervisors, middle and upper management
- Believes completely in quality circles as a method of developing employee potential
- Promotes, implements, operates, and manages a quality circle program
- Motivates, supports, and encourages people
- Sets standards and priorities
- Optionally participates in steering committee activities

---

have the personality to "get out and mingle with the people." The facilitator should believe in the quality circle concept and have the ability to instill the same enthusiasm in leaders and members.

The facilitator is normally responsible for a number of quality circles, and in large organizations with an extensive program, several facilitators may report to a senior facilitator. In the latter case, the senior facilitator should sit on the steering committee.

### Circle Leader

The circle leader is responsible for the operation of one quality circle and works closely with the facilitator in establishing the circle and training the members. The circle leader has essentially the same training as the facilitator without the same span of control. In fact, the facilitator and circle leader may work closely in member training and collaborate during the sessions.

The circle leader will frequently be the group's supervisor, although this need not necessarily be the case. Initially, it is a good idea to start with the supervisor as leader—provided that the members accept the arrangement. Usually, the supervisor has the breadth of experience to insure that the circle heads in the right direction. Later on, the supervisor can continue as leader or the group can elect one of the other members as the leader. In any case, the supervisor should participate in the circle's operation. Another consideration, of course, is that the supervisor is more likely to get ideas accepted by top management because a certain level of credibility already exists.

One approach to establishing the circle leader is to have the supervisor start out and, after a period of circle operation, appoint a member as assistant leader. Then, over time, the assistant can evolve into the full leader. It is also possible to elect the initial leader from within the group, but this approach tends to leave the supervisor dangling with nothing to do.

Table 3.6 gives a representative set of quality circle leader functions. Clearly, all items are significant, but perhaps the most important function is that the leader is responsible for the operation of the circle. Problem selection, analysis, data collection, participation and discussion, and finally problem solving are all under direct control of the quality circle leader. When participation wanes through a lack of challenging ideas, the leader initiates brainstorming to generate new ones. When members experience difficulty in communicating and getting their contributions recognized, the leader assists through encouragement and diplomacy. Finally, the leader is responsible for winding up a topic and for fostering its acceptance by management.

A circle leader may also participate in a leader circle that has the objective of applying participative management to the quality circle program itself. It is here that leader methodology is exchanged between circle leaders. Some of the topics that frequently arise are policies, codes of conduct, organization, scheduling, and training.

Several procedural items can assist in making a quality circle program run smoothly from the standpoint of the circle leaders:

- Start and end meetings on time.
- Plan the meetings in advance.
- Prepare, distribute, and use an agenda.
- Summarize meetings for the next session.
- Document results.
- Use a critique session to enhance performance.
- Control the discussion process so members stay on the right track and work towards an objective of problem identification, analysis, or solution.

In addition, the leader must train members, and the training sessions occupy the first few circle meetings. After the methodology is presented, meetings evolve into a problem-solving modality from a training program. While the responsibility for the smooth and effective operation of a quality circle is the circle leader's, the facilitator provides support as long as it is needed. The facilitator will attend early meetings, but this form of participation diminishes over time.

## Table 3.6  Quality Circle Leader Functions

- Trains members with assistance from the facilitator as needed
- Is responsible for circle activities
- Is responsible for the operation of the circle
- Assists with circle reporting and management presentations
- Follows up on action items
- Shows interest, enthusiasm, and support of the circle
- Prepares the agenda for circle meetings
- Enforces policies and rules of conduct
- Assists in analysis and problem solving

### Circle Members

The objective of the quality circle philosophy is to enlist the participation of circle members, drawn from an object department. Even though membership is voluntary and a certain amount of freedom exists for members to join, leave, and then rejoin a circle, there should be genuine effort on the part of management, the union, the facilitator, and the circle leader to encourage total participation. As a "person-building" technique, the benefits of quality circle participation should be available to all employees. Nowadays, employees are concerned with career paths. Well, a career path without participation management experience is like engineering without mathematics. In short, you can't get along without it.

Quality circle members are trained by their circle leader—often assisted by the facilitator. Members take training to learn the techniques and to gain familiarity with circle operations and group dynamics. Towards the end of the training period, the group attacks real problems in the workplace. Initially, quality circle problems are proposed by management, the facilitator, and the circle leader. Later, members propose problems on their own.

Table 3.7 summarizes the functions of the circle members. These functions are meaningful if the constitution of the group encourages participation on the part of members. Members should all do the same kind of work or at least work on the same problem. Quality circles formed as a combination of specialists—such as electricians, machinists, and plumbers—often discourage participation because some specialists are invariably observing while others are working on a problem. In general, it is best if members of a given quality circle

## QUALITY CIRCLE PRINCIPLES 53

**Table 3.7 Circle Member Functions**

- Attend periodic meetings
- Adhere to circle policies and code of conduct
- Learn and apply quality circle methods
- Participates in problem identification, analysis, and solution
- Participate in management presentations
- Encourage participation of other members and nonmembers

report to the same supervisor. When specialists are needed during problem solving, they can be brought in by the facilitator.

A famous baseball personality is quoted as saying "The game's not over until it's over." In a similar vein, the key to participative management is participation. Accordingly, members should

- Attend all meetings
- Actively engage in solving problems
- Follow the code of conduct

The *code of conduct* for circle members, given in Table 3.8, is nothing but common sense and ordinary courtesy. It helps to have the code contained in the training materials.

## OPERATION OF A QUALITY CIRCLE

There are two key aspects of quality circle operation: training and problem solving. During the start-up phase of quality circle operation, training is the main activity and the first two or three sessions are devoted exclusively to that subject. After problem solving has begun, however, training proceeds on a topical basis as new techniques become necessary to resolve new situations.

Problems are suggested by management, the facilitator, or the circle leader, or through member participation. The last case normally takes the form of a brainstorming session or an idea thought of by a member and communicated to the group. Realistically, most problems attacked by a quality circle are of the ordinary variety commonly associated with the everyday workplace. Only rarely do quality circle problems approach interdepartmental affairs or affect the organization as a whole. Yet, small problems invariably lead to big troubles, especially if they are compounded.

## Table 3.8  Code of Conduct for Circle Members

- Each member should attend all meetings
- Each member should participate
- Members may criticize ideas but not people
- Each member is free to express ideas or make suggestions
- Members should listen to other persons' contributions
- Members should work on a group project

The operation of a quality circle takes place through the following steps:

- Problem identification
- Problem selection (by members)
- Problem solving (by members)
- Recommendation to management
- Management review of recommendation
- Decision by management

The problem-solving phase is commonly augmented by specialists and with information from other areas of the organization.

While the steps appear to be self-evident from their names, the processes inherent in them may be a bit more complicated. For example, a typical quality problem experienced in manufacturing is the production of parts that are out of tolerance. Call this a *result problem*. There usually are many causes—each of which could be a potential problem to be solved by circle members. Data collection and data analysis may be necessary to identify the primary causes. Only then do the members select a *cause problem* to solve. Usually, the major cause problem is solved by the circle; however, this is not always possible. It is conceivable that the major cause problem is interdepartmental or organization-wide. Systems problems of this type require more than a quality circle—perhaps a task force. Also, some problems are industry-wide and are accepted by all persons involved. A quality circle would then attack an appropriate subproblem for solution.

In spite of the wide diversity of problem situations, an analysis procedure seems to be the most amenable to routinization and common to one or more of the phases. The following steps to the analysis procedure have been proposed in a variety of forms:

- Plan (P)
- Collect data (C)
- Analyze data (A)
- Draw conclusions (D)

The analysis procedure is the subject of Chapter 4.

Once a cause to a result problem has been identified, it becomes a problem in and of itself. Circle members, because of their experience with the subject matter, may have an immediate solution. More often than not, however, there may be different and possibly conflicting solutions. Discussion will ensue and eventually a consensus will be reached as to the most appropriate course of action. An implementation plan should be prepared at this point and the solution should be tested before a decision is made by the circle members to make a recommendation to management. Clearly, the P-C-A-D procedure may apply in this case, as well.

The recommendation to management is made through a management presentation and possibly a written report or a description in some form of the problem statement and recommended solution.

The management presentation is no small undertaking. The setting up of the meeting arrangements, per se, by the facilitator is only the beginning. Charts—and possibly slides—must be prepared, and several walk-throughs of the presentation scenario may be needed. The presentation should not be made to the steering committee but to the manager to whom the circle leader reports. Clearly, higher-level management may sit in—and this is obviously recommended. However, the review process should parallel the formal organization structure. Bypassing the normal control structure could easily lead to negative results.

The management review and eventual decision should take place in a reasonable time period. Actually, it is preferable to make the time of the review period a matter of policy. Otherwise, the circle may not be responsive to the next problem that surfaces.

## VARIETIES OF QUALITY CIRCLES

The dynamics of quality circle activities frequently leads to problems or subproblems that are either too big, too small, or too specialized for the group. In some cases, a major problem is caused by several subproblems, each of which is best handled by a select group under the leadership of a knowledgeable member. In other cases, a problem may be too small for a 12-person group to attack and it can be assigned to a subgroup. In a sense, "too many cooks. . . ." Not only do minicircles increase the sense of participation among members, but they also permit a sharper focus on a problem than is possible with a larger group.

In an analogous vein, quality circles also grow in experience, in competence, and in their ability to attempt more sophisticated problems. Often, problems span different shops—each of which has an ongoing quality circle. When problems span shops, departments, and offices, *joint quality circles* can be formed to solve common problems. Normally, joint quality circles involve shops one of which precedes the other in a manufacturing or commercial process.

As quality circle members become familiar with more analysis techniques and gain experience in the problem-solving process, mini and joint quality circles are a natural consequence. *Leader circles* are commonplace and even *facilitator circles* have been initiated.

## SUCCESS ELEMENTS

The elements that lead to a successful quality circle program are integral to many of the topics covered previously, such as voluntary participation and management support. In fact, dozens of books exist on motivation, human relations, communication, learning, and management. Nevertheless, there are ten critical success factors that stand out and should be seriously considered before a quality circle plan is put into effect. Clearly, each organization, or even each individual, may view them differently, but overall, they serve as a "success checklist:"

- Effective leadership
- Attention to rewards and recognition
- Linkage to the suggestion system
- Attention to group processes
- Adequate training
- Realistic goals
- Well-defined roles and expectations
- Promotional activities
- Efficient record keeping
- Careful measurement and testing

The ten critical success factors can be used by top management to oversee the quality circle program and should definitely be used by the steering committee and facilitators to gauge progress. Attention to the success factors coupled with an ongoing plan for monitoring, assessment, and control will lead to a successful program.

## SUMMARY

A quality circle is a small group of workers who meet regularly on a voluntary basis to analyze problems and recommend solutions to management. The organizational sphere of activity of quality circle participants is the area in which they work. The basic philosophy underlying quality circles is that quality awareness through participative management can not only identify problem situations but can also assist management in solving them. Some quality circle attributes are

- Participation is voluntary.
- A quality circle is a small group ranging from 4 to 12 people.
- Quality circle members do similar work.
- Quality circles meet regularly.
- Each quality circle has a formal leader.
- A quality circle program in an organization has one or more facilitators who guide the program.
- A management steering committee establishes objectives, policies, and guidelines for the program.

Each participant in a quality circle program receives adequate training to enhance his or her involvement in quality circle activities.

No organization change is needed to implement a quality circle program, but a successful program will most certainly change the organization. The key participants in a program are

- Steering committee
- Facilitator
- Circle leader
- Circle members

All participants, other than the facilitator, are drawn from the existing orce. The facilitator is a new position that may be staffed from manufacturing, industrial engineering, or the human resources department.

The basis for quality circle policy formulation is the specification of purposes and, eventually, objectives for the program. Key purposes are

- To enhance the capability of the enterprise
- To provide motivational factors
- To fully utilize and expand the capabilities of individuals

Based on these purposes, five universal objectives of a quality circle program are

- To improve productivity and quality
- To improve communications
- To enhance cooperative behavior
- To provide an efficient operation
- To enhance morale

The two main ways that quality circles are formed are as employees from the same work unit, but not necessarily performing the same task, and as employees performing the same task, such as word processing, in different organizational units. Accordingly, the major elements of a quality circle policy set are

- Employee participation
- Management support
- Management participation
- Quality circle activities
- Organization
- Restrictions

Clearly, each major element includes several policy statements.

The implementation of quality circles can proceed on a top-down or bottom-up basis. Top-down implementation refers to a total program initiated by top management and affecting the whole organization. Bottom-up implementation refers to the pilot project approach that is expanded as conditions and needs become evident. In either case, a successful program requires a sponsor to foster the concepts. The steps in quality circle implementation are well defined.

The operation of a quality circle takes place through the following steps:

- Problem identification
- Problem selection (by members)
- Problem solving (by members)
- Recommendation to management
- Management review of recommendation
- Decision by management

The problem-solving phase is commonly augmented by specialists and with information from other areas of the organization. In spite of the diversity of problem situations, an analysis procedure inherent in several of the above steps seems to be the most amenable to routinization. The following steps to the analysis procedure have been proposed in a variety of forms:

1. Plan (P)
2. Collect data (C)
3. Analyze data (A)
4. Draw conclusions (D)

In general, two types of problems exist: result problems and cause problems. A quality circle solves a result problem, such as excessive operating costs, by analyzing it and dividing it into subproblems, such as telephone utilization and outside services, through quality circle techniques and then solving major subproblems through quality circle activities.

The dynamics of quality circle activities frequently leads to problems or subproblems that are either too big, too small, or too specialized for the group. Thus, major problems can be divided into subproblems to be handled by minicircles, and problems that span shops or departments can be handled by joint quality circles. Other concepts include leader and facilitator circles.

Success elements for quality circles have been identified and can be used by top management to oversee the programs and should be used by the steering committee and the facilitators as checklists.

# 4 QUALITY CIRCLE METHODS

## INTRODUCTION

Some organizations with long experience in participative management, in management development/organizational development programs, and in quality assurance can use the top-down approach to quality circle implementation and have the resources to buffer the costs until positive results are received. Not everyone is convinced, however, that a quality circle program will be successful in his or her organization. In this case, the most prudent approach is to initiate a pilot project and then measure the results. If results are favorable, then the program can be extended into other areas. Otherwise, the program can be discontinued or initiated in another area. On the other hand, there could be an even more important reason to measure results from quality circle activities. An effective measurement program can justify continuation of a quality circle program when political winds change and a new administration is not convinced of a program's viability. At the quality circle level, measurement and analysis techniques are used for problem identification, problem analysis, and problem solution, and to support recommendations to management. This chapter covers measurement, data collection, and methods that are applied to data collected. In the last category, statistics, cause-and-effect diagrams, and Pareto diagrams are introduced. Brainstorming is covered as a method for problem identification and analysis.

## MEASUREMENT

Measurement to support, justify, and analyze the operations of a quality circle program is essentially no different from any other type of measurement. Because quality circle members, as a general rule, want to exhibit good results, they characteristically have a good attitude towards measurement. In the domain of quality circles, measurements can be grouped into three categories: quality indicators, cost analysis, and attitude indicators.

**Quality Indicators**

Clearly, *quality indicators* differ between production shops and service industries. Nevertheless, a similar mode of thought exists between the two areas, as evidenced by the representative set of quality indicators given in Table 4.1. However, it should be mentioned that indicators are not always obvious. In a data processing shop, for example, errors due to mounting the wrong tape were counted and analyzed in a daily meeting. Shrewd computer operators, however,

Table 4.1  Representative Quality Indicators

Billing errors
Breakdowns
Contract errors
Customer complaints
Customer rejections
Defects
Defects per unit of work
Delays due to errors
Design changes
Later deliveries
Lawsuits
Methods improvement
Process changes
Purchase order changes
Recalls
Repeat purchasing
Repeated errors
Repeated failures
Reruns
Review-based actions
Rework costs
Scrap levels
Shortages
Spoilage
Time lost due to breakdowns
Tool rework
Warranty claims
Yield rate

determined that reruns of the same job were not recorded. So when an incorrect tape was mounted, they allowed the job to run to completion, and then reran it with the correct tape. This quality system literally wasted thousands of hours of valuable computer time.

One way of looking at quality indicators is to draw a chart, as shown in Figure 4.1. In Figure 4.1(a), the causes are organized with regard to frequency, which gives one a picture of the quality situation. The most frequent cause is placed first, the second most frequent cause is placed next, and so forth. In terms of cost, Figure 4.1(b), quite a different result is obtained. In this case, causes are arranged according to the cost impact on the organization.

In many organizations, workers can easily cover up errors and there is no means of measuring reruns, recalls, and customer inconvenience—except at the bottom line. Through quality circle activities, many sources of poor quality are uncovered through the normal processes of group dynamics. Thus, it is realistic to contend that quality circles work *with* measurement techniques, rather than stating that one of them is dependent upon the others.

### Cost Analysis

Cost figures are the language of production shops and service industries. Therefore, any cost reduction achieved through quality circle activities is relative to the corresponding cost level. One measure that is commonly used is the savings-to-cost ratio. When an improvement is recommended in a procedure or process, the costs and savings are projected into the future to give a realistic picture of the value of the suggestion.

All quality improvements can be translated into either cost reductions or productivity gains. Production people have a variety of methods of projecting production and cost data into the future to estimate the payback or improved cash flow of a proposed scheme. Perhaps the most powerful and most easily understood of these methods is the learning curve—commonly associated with motor learning in humans. The key to the learning curve, as applied to production, is that the rate of improvement (or *gain*) in a process is sufficiently well defined to be predictable. This is the case because of familiarization and experience of personnel, adjustments and fine tuning of equipment, enhanced human-machine interaction, enhanced

QUALITY CIRCLE METHODS 63

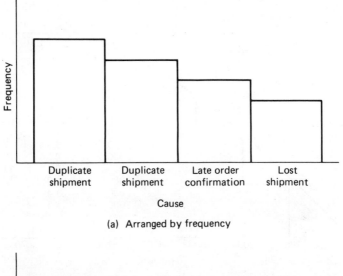

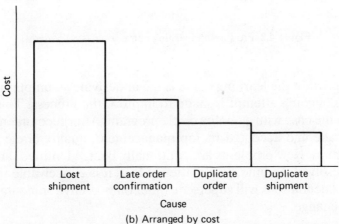

**Figure 4.1** Analysis of quality indicators, (a) arranged by frequency; (b) arranged by cost.

processing flow, and improved design methods so that products are designed to be manufactured in addition to being designed for hard use.

The learning curve implies that production tends to improve by a constant percentage each time the production doubles. Figure 4.2 depicts a learning curve at a percentage of 80 percent. This means that the second 1000 units require only 80 percent of the resources required by the first 1000 units, and so forth.

## 64  QUALITY CIRCLES

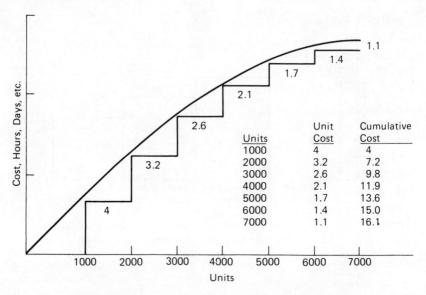

**Figure 4.2** Production learning curve (rate = 80 percent).

Implicit in the learning curve is the underlying assumption that an active ongoing attempt is made to improve the process. This is precisely the case with a quality circle program. After recommendations are made and accepted by top management, quality circle activity continues. New problems are continually being identified, analyzed, and resolved. In time, as a matter of fact, it is conceivable that new recommendations will supersede older ones as environmental conditions change.

### Attitude Indicators

Many organizations are adept at measuring production and service data but have limited experience with the measurement of employee attitudes. Although the items to be measured seem to be obvious, once they are mentioned, many people simply have not thought about them. This is probably the case because they haven't had to do it or didn't think it was important. Well, employee attitudes are important and a quality circle program can significantly improve employee motivation through participative management. Attitudes can be improved even if they are already good. There is no need to describe the following attitude indicators:

- Employee absences
- Tardiness
- Grievances
- Turnover
- Attitude surveys
- Direct observation

For absences, tardiness, grievances, and turnover, personnel records are a satisfactory source of information. Attitude surveys and direct observation require "human resource" training and experience, although some organizations may want to handle the measurements informally.

Measurement, in the sense used in this section, refers to the evaluation of a quality circle program. General guidelines on the specifications for the measurements of a quality circle program are normally determined by the organization's strategic plan.

## BRAINSTORMING

In the dynamics of quality circle activities, ideas are needed for three situations:

- Identifying problems
- Uncovering probable causes
- Developing viable solutions

A commonly used technique to stimulate ideas in cases such as these is brainstorming.

### Basic Concept

*Brainstorming* is a group technique for generating ideas. The basic philosophy underlying brainstorming is that a response by one person seems to stimulate the creativity of other participants. Often, ideas are held back by a person for a variety of reasons:

- The idea doesn't seem relevant.
- The person is cautious in presenting an idea for fear of being ridiculed.
- The person is not exactly sure that his or her input is wanted.
- The idea is dormant and the right stimulus to cause it to surface is needed.

The reason that brainstorming is successful with quality circle activities is that the group is involved with the work environment that they know best: their own.

Brainstorming is a means of generating the maximum number of ideas on a topic. The brainstorming procedure works as follows:

- The topic is introduced and the scope of the desired input is clearly stated.
- The guidelines of brainstorming are briefly reviewed, just to be sure. (This topic will have been covered in the circle member's training program.)
- Each participant, in turn, is asked for an idea.
- The ideas are listed on a flip chart or a chalkboard by the circle leader. (The leader may have to summarize an idea, with the originator's agreement.)
- Members are solicited for ideas in rotation, until all input is exhausted.

After as many relevant ideas as possible are generated, the circle leader, with the help of the members, goes through an analysis phase to identify the top ideas.

In some cases, an insufficient number of ideas is generated on a topic and the session can be suspended until the next meeting. An incubation period is sometimes beneficial since a successful session is dependent upon the fact that the members are actively familiar with the topic and the proper stimuli are present. By "sleeping on a problem" for a week or two, members have a chance to "wake up" to a problem, cause, or solution.

### Guidelines

The adherence to recognized guidelines can greatly increase the yield of a brainstorming session. The worst thing that can happen is to have members "clam up," because they or someone else has been criticized or laughed at. Also, it may take a considerable amount of courage for some people to give ideas, and these are precisely the ideas that are desired.

The guidelines are simple but effective:

- Each member may offer only one idea per turn.
- The ideas are not discussed. Clarification, if necessary, is permitted and the idea is then recorded in summarized form.

- Criticism and judgment are not permitted.
- Informally, good naturedness, laughter, and free wheeling are encouraged. Wild off-beat ideas can trigger a really useful concept. Members are more creative when the atmosphere is not tense and formal.
- Strive for quantity of ideas. Quality can come later.
- Encourage improvements to ideas and combinations of ideas.

The biggest problem that the circle leader has is to keep the session on the subject. In spite of its informality, a brainstorming session is not for social purposes. The circle leader must reinforce, through his or her actions, that the process is deadly serious.

The members should also be encouraged not to allow their egos to become involved with an idea. Because of an improvement upon an idea or a combination of them, a team member can easily get the impression that his or her idea was effectively "stolen." Members should be constantly reminded that quality circle activity is a team process, and also that other members are well aware of where an idea came from—even if they do not say it.

### Analysis of Ideas

Ideas must be discussed, analyzed, and judged. Clearly, "off the wall" ideas must be discarded and sound ideas should be reinforced.

The task of narrowing down the list of ideas is done with a simple voting method. Team members are allowed to vote on each idea in turn, and the ideas with the most votes enter into a discussion phase.

At this point, team members can concentrate their attention on a smaller number of items, where thorough and exhaustive discussion is needed and encouraged. After the discussion period, which may extend to two or more sessions, a final vote is taken on the remaining ideas.

Afterwards, study groups, minicircles, joint quality circles, and data collection and analysis are performed prior to the task of preparing a final recommendation to management.

## DATA COLLECTION

In manufacturing and other forms of commerce, a large amount of data is collected to form the basis for decisions and actions. Reasons for collecting data can be conveniently grouped into the following categories:

- To assist in understanding the situation
- To serve as a basis for analysis of cause and effect
- To provide a basis for regulating the process
- To support statistical quality control
- To establish a basis for financial analysis

In most cases, the reporting aspect of data collection is taken for granted, and in many forward-looking organizations, only exception data is reported in order to cut down on the volume of paper required for decision making and control. It is assumed throughout that the data collected represents the facts and that the analysis is appropriate to the problem domain. Accordingly, sampling and statistical analysis, respectively, constitute the foundation of data collection.

There are essentially two aspects of data collection that must be considered:

- Recording of data
- Presentation of data

The *recording* phase accumulates raw data, and the *method of presentation* is used to display the data in a form that is readily understandable. Clearly, these are both broad topics that can only be introduced here. Moreover, quality circle activities do not in general require advanced methods, and straightforward and easily understood techniques for recording and presentation are more than satisfactory for most applications.

### Recording of Collected Data

Data must be recorded in order for it to be useful for quality circle activities. Four methods are used frequently enough in this area to be mentioned:

- A checklist
- A drawing
- A check sheet
- A computer printout

## QUALITY CIRCLE METHODS 69

The methods are sometimes used in combination. A *checklist* is simply a list of actions performed or items to be inspected. A checklist specifies two items of information: a tabulation of entries (i.e., actions or items) and an implicit order in which the entries should be considered. As an example, a checklist could reflect an inspection record wherein various conditions are either accepted or rejected. Another type of checklist could simply denote the presence of a certain defect or failure.

Another means of recording data, especially in the quality assurance area, is to denote the existence of a defect on an artistic or mechanical *drawing* for the product. This form of recording could be used to represent flaws in sheet metal or paint work, such that the area of greatest susceptibility is signified by the density of marks.

The most frequently used form of recording is the *check sheet*, as suggested by Figure 4.3. On one axis, an event—such as a quality problem—is outlined. On the other axis, a domain—such as a product model—is represented. Thus, occurrences of various events for specified products can be signified by a tabulation mark. The key advantage of this form of recording is that the marks can be totaled by event and also by product. The check sheet method of recording also serves as an excellent input to the presentation phase.

The final method of recording, covered here, is the computer

|  |  | Product | | | | |
|---|---|---|---|---|---|---|
|  |  | Model A | Model B | Model C | Model D | Total by Quality Problem |
| Quality Problem | Door Molding |  |  |  |  | 11 |
|  | Sheetmetal |  |  |  |  | 12 |
|  | Paint Finish |  |  |  |  | 25 |
|  | Alignment |  |  |  |  | 6 |
|  | Driveability |  |  |  |  | 22 |
|  | Total by Product | 25 | 13 | 25 | 13 | 76 |

**Figure 4.3** Representative check sheet for data recording.

printout. Many processes in modern business are controlled by computers. Process control computers, various forms of automation, data entry stations, and banking terminals are typical examples. Data recording is a by-product of operations in this class, and errors of various kinds are tabulated, summarized, and reported for management review. This is an excellent source of information provided that the computer output can be produced in a useful form.

Finally, some thought should be given to the process of recording. Some of the usual considerations are the time period for collection, the kind of data to be collected, and form design, in the case of checklists, drawings, and check sheets. Sampling is a key issue in data collection, and many quality circle groups rely on expert advice from the quality control or industrial engineering departments in this regard.

## Presentation of Collected Data

The most common method of presentation is with a graph that gives a visual model of the concept being presented. An arbitrary distinction is made here between the presentation of collected data and the presentation of analysis data. In the former case, the following concepts are presented:

- Line graphs
- Histograms
- Pie charts
- Control charts

In the latter case, two methods prevail:

- Pareto diagrams
- Cause-and-effect diagrams

The Pareto and cause-and-effect diagrams are introduced in the next section. In the final section, the methods are combined as they would be in quality circle analysis.

A *line graph* is normally used to represent collected data as it is summarized over time, as suggested by Figure 4.4. A line graph does not automatically engender any corrected action on the part of employees or a quality circle because causes are not obvious from data. In fact, a line graph may be generated in response to a quality circle's request for data. It follows that brainstorming or another analysis technique would be used to identify causes and possible solutions.

Closely related to a line graph is the *control chart,* shown in Figure 4.5, on which upper and lower control limits are drawn. Usually, an average value of a variable, such as tolerance, is recorded versus time.

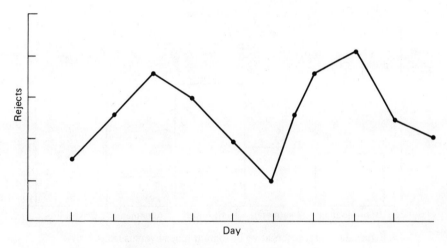

**Figure 4.4** A representative line graph.

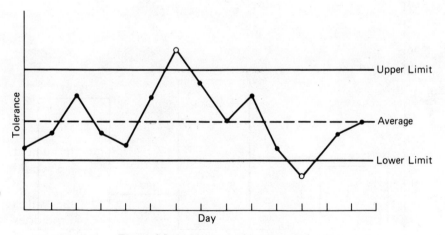

**Figure 4.5** A representative control chart.

**72  QUALITY CIRCLES**

When a periodic value falls outside of the specified range, an investigation into the causes is automatically triggered. As with line graphs, the application of brainstorming or another analysis technique is needed to identify causes and possible solutions.

A *histogram* is a visual diagram in which frequencies are represented by the height of a vertical bar, as depicted in Figure 4.6. A slight variation to the frequency histogram is the percentage histogram, as shown in Figure 4.7, where percentage instead of frequencies is

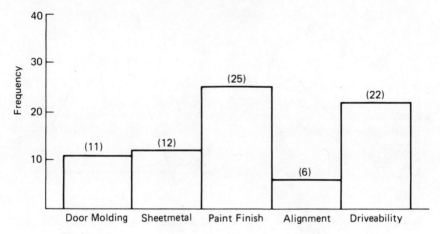

**Figure 4.6**  A representative histogram (cause versus frequency).

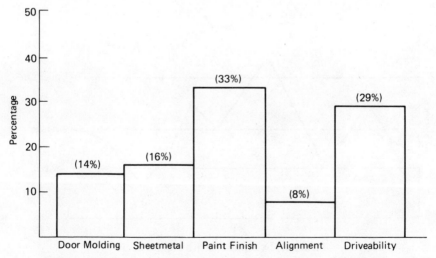

**Figure 4.7**  A representative histogram (cause versus percentage).

depicted. The shape is the same in either case, so each is an analog of the other. A histogram is used in quality circle activities to identify problems for group solution. In Figure 4.6, for example, the major quality problems are "paint finish" and "driveability." Clearly, if these major problems were resolved, the quality picture would be considerably different.

A final method for presenting collected data is the *pie chart*, shown in Figure 4.8. A pie chart is yet another technique for representing frequency and percentage data and is directly analogous to the histogram technique. Selection among the various methods is largely arbitrary.

### Advanced Methods

Additional methods for presenting collected data are scatter diagrams, project charts, and pictograms. Applicable statistical techniques are stratification and correlation. A basic book on applied statistics would be useful for reviewing these methods.

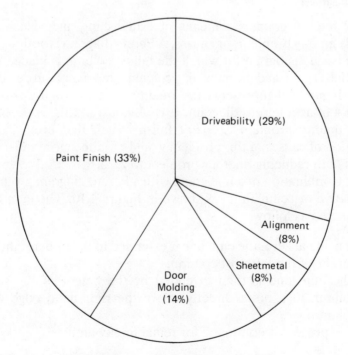

**Figure 4.8** A representative pie chart.

## Discussion

Quality circle methods are relatively simple and straightforward because they reflect an attempt to match methodology with the experience level of the participants. Training is problematical because it is impossible to devote more than an hour to a topic in any one session, and this amounts to no more than a brief glimpse at theory, procedures, examples, an exercise, and review questions. Advanced techniques are within the domain of systems analysis, industrial engineers, and quality control specialists.

## DATA ANALYSIS

Data analysis techniques facilitate the process of identifying problems, causes, and solutions and presenting recommendations to management. Two basic methods are used in quality circle work: Pareto diagrams and cause-and-effect diagrams. Cause-and-effect diagrams are further classed as "basic cause and effect" and "process cause and effect."

### Pareto Diagrams

A Pareto diagram is a means of establishing and visualizing a priority among problems or causes. A Pareto diagram is nothing more than a basic histogram for which the tallest column is placed to the left in the diagram and the remaining columns are arranged in descending order. Figure 4.9 represents the quality data (originally shown in Figure 4.6) as a Pareto diagram. It is obvious, in this case, that the major quality problem is "paint finish," such that even a partial reduction of causes of this problem would produce more satisfactory results than reducing another problem to zero.

The combination of a line graph with a Pareto diagram produces a cumulative or "cum" line, as shown in Figure 4.10. The cum line is constructed as follows:

- Starting at zero, the cum line is extended to the upper right-hand corner of the leftmost column.
- The cum line is then extended by the height of the second column to a point directly above the right-hand edge of the column.
- The process is continued for remaining columns.

The cum line is complete when it represents 100 percent of the cases.

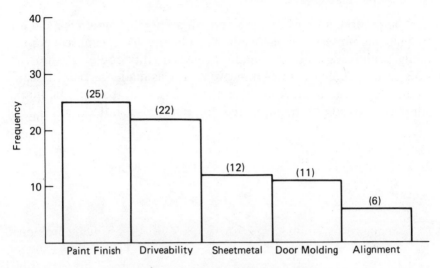

**Figure 4.9** A representative Pareto diagram.

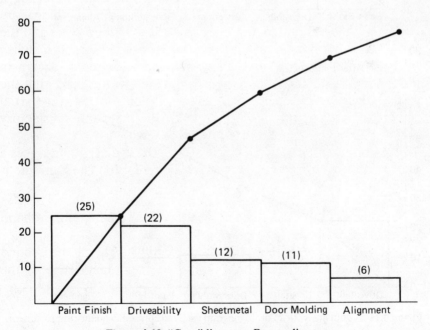

**Figure 4.10** "Cum" line on a Pareto diagram.

## 76 QUALITY CIRCLES

The combination of a cum line with a Pareto diagram is ideal for management presentations, as shown in Figure 4.11, because it graphically depicts a gain that is made through quality circle activities. In this case, a quality control improvement was made to "paint finish," reducing the number of problems from 25 to 10. It should be noted that the position of the improved column is also shifted.

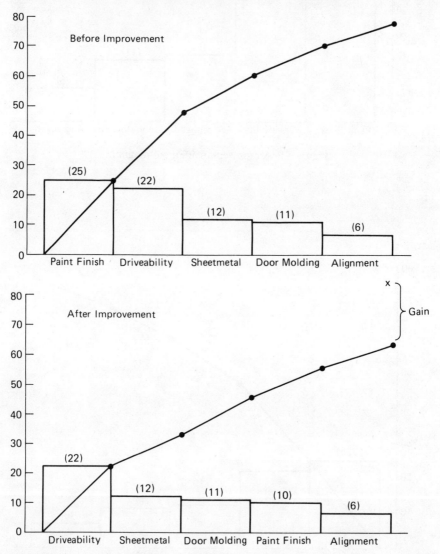

**Figure 4.11** A cum line with a Pareto diagram used to represent a quality improvement.

As noted previously, a variation to a Pareto diagram is to represent the "cost of quality," associated with various quality problems, instead of frequencies or percentages. This gives an alternate view of cost reduction and can shift priorities markedly.

### Basic Cause and Effect

When a problem (i.e., an effect) is known and its causes are not known, an excellent group technique for problem analysis is to construct a cause-and-effect diagram. Because of its visual effect, a diagram of this type looks like a fishbone, hence the name "fishbone diagram," or "Ishikawa diagram," after Professor Kaoru Ishikawa, who invented it.

The steps in the construction of a cause-and-effect diagram are given in Figure 4.12 and are delineated as follows:

- The problem to be analyzed is represented by a box with an arrow running into it from the left (step 1).
- The circle members, through group techniques, identify major causes, which are connected to the main arrow by slanting arrows and labeled (step 2).
- Minor causes are determined and connected by arrows to the major causes and labeled (step 3).

After the diagram has been completed, the members vote on the causes they feel are most important, as in brainstorming. The important causes thus identified are further studied for problem resolution.

The construction of a cause-and-effect diagram is a dynamic process that requires a high level of member interaction. A sparse diagram reflects a shallow knowledge of the process. A bushy diagram reflects a quality session in which the circle leader did not effectively control the group discussion.

### Process Cause and Effect

It is also useful to model a cause-and-effect diagram after the steps in a production process. This type of analysis is particularly useful for assembly-line problems.

The steps in the construction of a process cause-and-effect diagram are given in Figure 4.13 and are delineated as follows:

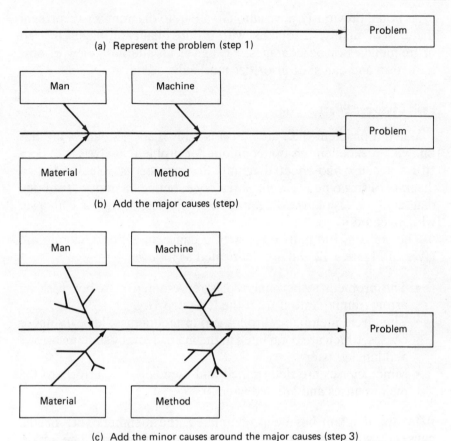

**Figure 4.12** Steps in the construction of a basic cause-and-effect diagram.

- The problem to be analyzed is represented by a box with an arrow running into it from the left (step 1).
- The steps in the production sequence are established by working forward, backward, or in both directions (step 2).
- Major and minor causes are determined and are connected by arrows to the boxes representing the production sequence (step 3).

In this method, the major and minor causes are identified by brainstorming and, as before, quality circle members vote on the most important items for further analysis.

This type of diagram is easy to make and understand because it closely models the real process. A minor disadvantage is that the

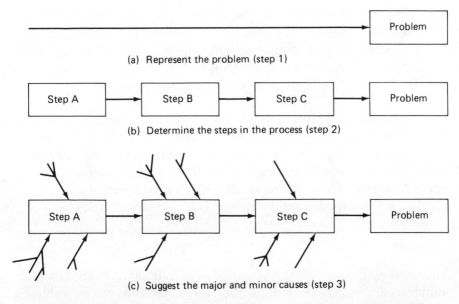

**Figure 4.13** Steps in the construction of a process cause-and-effect diagram.

same causes tend to appear as arrows attached to several steps in the sequence and the "bushiness" of the diagram tends to increase rapidly.

## APPLICATIVES

An *applicative* is the application of two or more data collection and analysis methods in succession to solve a quality circle problem. As an example of the concept, assume the existence of a complex manufacturing quality problem as suggested by Figures 4.3 and 4.6. A representative applicative for this problem is given as follows:

- The major quality problem is recognized. Major failures are listed.
- A check sheet is designed to collect the quality data.
- Production data is collected.
- A frequency or percentage histogram is constructed.
- From the histogram, a Pareto diagram is developed.
- The major problems are isolated.
- Process cause-and-effect diagrams are constructed through knowledge of the sequence of steps in the production procedure and the method of brainstorming.

- Circle members vote on the most important causes.
- Through small-group dynamics, including further use of brainstorming and possibly other methods, tentative resolutions for the various causes are identified.
- Recommendations are presented to management for approval.

This applicative is typical of the series of steps that would take place for a certain class of problems in a quality circle. Clearly, each problem is unique in some way, so a good grasp of the quality circle methods is important. Experience is useful for designing applicatives, but basic knowledge of the methods is absolutely necessary.

It follows that quality circle training is an ongoing activity.

## SUMMARY

The most prudent approach to quality circle implementation is to initiate a pilot project and measure the results. If results are favorable, then the program can be extended into other areas. Otherwise, the program can be discontinued or started in another area. There could be an even more important reason to measure results from quality circle activities. When political winds change and new management is not convinced of the value of quality circles, effective measurement can support and justify the continuation of the program. At the quality level, measurement and analysis techniques can be used for problem identification, problem analysis, and problem solution, and to support recommendations to management.

Measurements to support, justify, and analyze the operations of a quality circle program can be grouped into three categories: quality indicators, cost analysis, and attitude indicators. All quality improvements can be translated into either cost reductions or productivity gains. One of the most powerful and most easily understood methods of projecting production and cost data into the future to estimate the payback or improved cash flow of a proposed scheme is the learning curve. Quality circles work with measurement techniques and neither is specifically dependent upon the other. A quality circle program represents an ongoing activity where new problems are continually being identified, analyzed, and resolved. A quality circle program can also significantly improve employee motivation through participative management.

Brainstorming is a group technique for generating ideas and is used in quality circle activities for identifying problems, uncovering probable causes, and developing viable solutions.

In manufacturing and other forms of commerce, a large amount of data is collected to form the basis for decisions and actions. Reasons for collecting data can be grouped into a few categories:

- Understanding the situation
- Serving as a basis for cause-and-effect analysis
- Regulating the process
- Supporting quality control
- Performing financial analysis

Two aspects of data collection are of prime importance: recording and presentation. For quality circle activities, four frequently used recording methods are the checklist, the drawing, the check sheet, and the computer printout. The most commonly used methods for presenting collected data are line graphs, histograms, pie charts, and control charts.

Data analysis techniques facilitate the process of identifying problems, causes, and solutions and presenting recommendations to management. Two basic methods of data analysis are used in quality circle work: Pareto diagrams and cause-and-effect diagrams. Cause-and-effect diagrams are further classed as "basic cause and effect" and "process cause and effect." A Pareto diagram is a means of establishing and visualizing a priority among problems or causes. A Pareto diagram is a basic histogram for which the tallest column is placed to the left in the diagram and the remaining columns are arranged in descending order. The combination of a cumulative line with a Pareto diagram is ideal for management presentations. A basic cause-and-effect diagram takes the form of a fishbone with large and small arrows representing major and minor causes for problem resolution. A process cause-and-effect diagram is used to model a basic cause-and-effect diagram after the steps in a production process. Major and minor causes are then identified for subsequent resolution.

An applicative is the application of two or more data collection and analysis methods in succession to solve a quality circle problem. Experience is useful for designing applicatives, but a basic knowledge of the methods is necessary. Quality circle training is an ongoing activity.

# 5 STRATEGIC PLANNING FOR QUALITY CIRCLES

## INTRODUCTION

Quality circles are unique in that they combine a participative management technique with a historically structured working environment. The technique is, of course, quality circles implemented as a small-group activity and the working environment is the classical production shop. Clearly, the notion of a production shop can encompass workers in manufacturing, banking, the office, and the total domain of the service industry. In short, structured tasks are characteristically performed by production workers—such as assembly workers, bank tellers, or inventory clerks—as compared to knowledge workers—such as managers, scientists, engineers, or analysts—who characteristically perform unstructured tasks.

Because diverse groups in a business enterprise are impacted by a quality circle program and the implications can be far-reaching, planning is particularly significant.

This chapter covers the key areas of strategic planning for quality circles, selection of projects, justification, and acceptance. Collectively, the information presented here can be used as a guideline to the development of a strategic plan for a quality circle pilot project and subsequently an overall quality and functional plan.

## SYSTEMS APPROACH TO PLANNING

It is important to recognize that a variety of group plans effectively feed into the total business plan, as conceptualized in Figure 5.1. Because of the enthusiasm that is commonly associated with quality circles, the claims for success can easily lead to overly optimistic expectations. It is particularly significant that the "quality" component of the plan be developed on a firm foundation.

The systems approach to planning is an attempt to put as much precision into the planning process as possible. The interrelationship

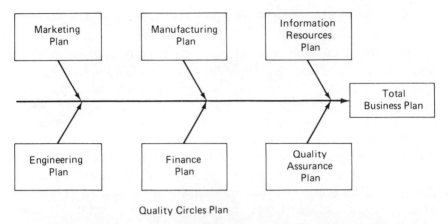

**Figure 5.1** Conceptual "fishbone diagram" for the total business plan.

between groups is a dynamic process that is continually being restructured as business conditions change. Therefore, management should insure that the most appropriate concept is applied to the most pertinent problem at the most opportune time.

Systems analysis for the purpose of planning should proceed along at least four dimensions:

*Congruence*—the most appropriate concept is used at the proper time.
*Integrity*—the approach provides a complete solution to the problem as it is viewed by the organization.
*Auditability*—the process can be demonstrated to provide the results specified.
*Controllability*—the operation of the procedure can be effectively controlled by management.

In fact, a good strategic plan should specifically address each of the dimensions.

As a concept, quality circles are observable, capable of being described in terms of procedures, inputs, and outputs, and are amenable to experimental manipulation to verify (or support) their congruence, integrity, auditability, and controllability.

*Congruence:* The dimension of congruence is related to the logic of systems analysis and the degree to which it reflects the relative capital-labor ratio. The methods of quality circles may have too much overhead, in terms of training, planning, involvement, and so forth, to be applied to certain small-sized management or production problems. On the other hand, the concept may be too weak to even assist in solving large-scale interdepartmental and organizational problems. Quality circles should be used *only* when they are the most appropriate method.

*Integrity:* The dimension of integrity is related to the degree to which a method or technique performs according to its specifications. In quality circles, there is an expected payoff in terms of improvements in costs, scheduling, quality, employee attitudes, and so forth. A quality circle is not a social gathering, and management should be able to predict, to some degree anyway, what results can be expected of a quality circle program.

*Auditability:* The dimension of auditability refers to the demonstrability of expected results through observation of actual circle operation and the resulting management recommendations. In short, a quality circle program must produce results and those results should end up on the proverbial bottom line. The results should be directly observable and measurable and not be hidden in a myriad of production and/or operational facts and figures.

*Controllability:* This dimension determines the extent to which management can control a quality circle program through directives, problem suggestions, the reward system, training, and actual management participation. Management should have direct or indirect control over the problems on which a quality circle works and should have a clear policy on how quality circle recommendations to management are handled. Clearly, management does not want to be faced with the situation of being coerced into accepting a recommended solution for fear of creating an adverse morale or attitude problem. There are two eventualities that management absolutely should avoid: (1) having the tail wag the dog; and (2) upsetting the apple cart through quality circles.

In short, a quality circle program should be describable and controllable, and the overall responses to business conditions should be predictable.

## STRATEGY CONCEPTS

A *strategy* is the means that management chooses to use an organization's resources to reach its objectives. Thus, a strategy is an implicit relationship between the organization, its objectives, its resources, and its environment.

A *statement* of strategy is a mechanism for focusing the attention of management planners on a topic and serves as a communications medium for management review, approval, and support. A statement of strategy includes, but is not limited to, the following topics:

- A comprehensive course of action
- Anticipated risks
- Organization dependencies
- External business environment and conditions
- Required resources
- Financial requirements, conditions, and projection
- Viable alternatives

In order to make a strategy statement, a strategic planner must be involved with quality circle concepts, as well as be knowledgeable of the organization's methods, history, planning, and politics.

Another important consideration is that the planner may not be a quality circle specialist, so some preparation may be required. The steps involved in developing a strategy are as follows:

- Understand the scope and nature of the subject matter, as well as its organizational implications.
- Describe the future environment.
- Identify the objectives and the alternative possible strategies.
- Set up criteria for selection of an optimum strategy.
- Select the preferred strategy.
- Prepare the strategy statement.
- Obtain strategy approval.
- Take steps to incorporate a functional strategy for quality circles into the strategic and operating plans of the organization.

It is important to recognize that a strategy is *not* a step-by-step plan for reaching a predetermined set of objectives, but rather is a metaplan. In this context, the term *metaplan* refers to the planning process about planning. Thus, a strategy paper should also delineate objectives and guidelines for the planning process itself, in addition to the preparation of operational plans and goals for the organization.

To sum up, therefore, the dimensions of a quality circle strategy are threefold:

- Directions and goals
- A guide for implementation
- A rationale for decision making

and the strategy deals primarily with: "How do we get there from here?" Effectively, a strategy provides relevant action for a future operational environment.

## STRATEGY AND PLANNING

The concept of a quality circle is not an end in itself. It is a method, albeit a general one, for organizational problems related to quality, performance, employee attitudes and motivation, scheduling, excessive costs, and so forth. The key question is not what should be done tomorrow, but rather, what should be done today to prepare for an uncertain future. A quality circle program may be implemented in a specific area, such as manufacturing or office automation, or it may involve the total organization. Nevertheless, an explicit decision is made to use quality circles. Organizations do not and should not employ the quality circle technique "on a whim" or because somebody likes the idea. The implementation of quality circles in your organization should be a direct response to an organization need.

### Contents of the Strategy

A quality circle strategy gives three things:

- Where we are (*Current position*)
- Where we are going (*goals*)
- How we get there (*direction*)

The *current position* is a specification of the equipment and supplies, already tested applications, the trained and knowledgeable people, and existing organizational problems that have a bearing on a quality circle plan. *Goals* are certainly dependent upon a particular enterprise but include factors such as better customer service, increased sales volume with the same head count, reduction of administrative expense, job enhancement, the establishment of new markets, improved quality, reduced costs, and more timely and effective decision making.

The *direction* (or "How do we get there?") is a major issue—in fact, it is the reason for a strategy in the first place. Direction needs policies and procedures in the following areas:

- Justification
- Implementation
- Employee acceptance
- Staffing and organization

The above areas form the basis for a strategic plan that gives the stages of tactical (or functional) planning for the enterprise.

### Strategic Plan

The strategic plan for a quality circle program covers three stages of work: preparation, development of a tactical plan, and utilization of the tactical plan.

The *preparatory work* is crucial because it sets the stage for success or failure. The history of planning in an enterprise should first be consulted because a quality circle will eventually become part of a larger planning entity. A high-level management commitment—or sponsor—is needed to kick off an effective project. Responsibility for a quality circle program and also for tactical planning should be assigned and a strategy group should be formed. Participation in any of the above activities need not be on a full-time basis, but the key objective is to specify basic objectives and output of the strategy sessions. Actually, an ad hoc task force to initiate a project, as outlined above, can be an effective course of action.

The *sponsor* is particularly significant because resources have to be obtained for a quality circle program even for planning itself, and the needed policies must be set and enforced.

In *developing the plan,* assumptions must clearly be made concerning the organizational structure, the staff, and the organizational implications. After gathering and analyzing information about the organization, a sequence of potential quality circle applications is selected. Objectives are required at this point, and they must minimally include opportunities, measurable goals, time frames, and any anticipated problems. The strategic plan, at this point, should be reviewed with local management and then be presented to top management.

In *utilizing the plan,* approval is requested for initial activities that include pilot projects, research efforts, and associated development work. A means of using the plan is also needed, which includes people, procedures, an employee feedback channel. The plan should also include an action item to be updated in line with the planning policies of the enterprise.

### Guidelines

When a strategy is outlined and a plan developed for a quality circle program, a few pertinent guidelines are helpful for increasing the chance of project success. The most important consideration is to concentrate resources in time and in place and focus these resources on the stated objectives. Second, it is not prudent to allocate resources unless there is a better-than-average chance of success. There are simply too many intangibles in participative management to insure immediate success, and employee acceptance can be problematical unless handled properly. It follows that proper consideration of employee reaction to a quality circle is an important aspect of strategic planning. Last, it goes almost without saying that you shouldn't compete for the organization's resources unless the conditions are favorable to the capacities of your organizational unit and also you shouldn't try to obtain these resources unless they will contribute to your organizational unit's strategic objectives.

## JUSTIFICATION

Enthusiasm over quality circles is contagious. The spirit of participative management, for example, is both exciting and interesting. It is unlikely that this enthusiasm will carry over to the executive suite, however, where seasoned executives will question the viability of the newly discovered source of productivity. Therefore, justification will be required—regardless of the credibility and access of the sponsor.

### Definition

Justification is the information presented to a decision maker to support an investment proposal. In general, there are two *major* reasons to provide justification:

1. To achieve agreement to put an investment proposal into an overall plan for the organization
2. To obtain a commitment of resources for an implementation project

In the area of quality circles, justification is notably difficult because hard money initially may be balanced against soft savings.

**Methodology**

Traditional methods of justification include almost any form of inductive reasoning. Some of the more noteworthy are

- Cost of participative management versus the old method
- Improved customer service
- The idea that there is no other way to do it
- Competitive advantage
- The reasoning that it is part of the cost of doing business
- Organization's "image"—to be at the leading edge
- Lower costs
- Improved quality
- Better employee attitudes and morale

With quality circles, a useful approach is to depict a labor cost curve with a flattened slope achieved by increasing the efficiency and effectiveness of the affected personnel. This is productivity.

**Productivity**

Productivity is a key benefit of the quality circle concept because it provides an increased quality and quantity of work, an increased span of control, more effective (i.e., more timely) work, and decreased turnover of key people. Thus, *productivity* can be more formally specified as the relationship between the output of a work environment to its input in labor and raw materials.

**Methods of Analysis**

Two methods of delineating productivity are the payback method and the cash flow method.

The *payback method* reflects costs that are displaced because of the results of quality circle activities. Reduced material requirements in a processing operation, initiated as a quality circle recommendation to management, is an example of a case where costs are actually decreased.

With the *cash flow method* of analysis, costs are not replaced by the newer management philosophy but the total output is greater for the same costs. A recommendation to reorganize the work area and promote team behavior is an example of how output could be increased without necessarily reducing costs.

In some cases, therefore, labor, equipment, and facilities costs are displaced and *cost savings* result. Typical examples are floor space for files, travel costs, postage costs, forms and supplies costs, time savings, and reduced materials. In other cases, *cost avoidance* takes place because there is decreased growth of staff, equipment, and facilities to accompany an increased output level.

## IMPLEMENTATION

The notion of *implementation* refers to a pilot project to establish the quality circle concept and then an expansion both horizontally and vertically within the organization. In this context, horizontal expansion refers to a proliferation of quality circles throughout the organization. Vertical expansion refers to an enhancement of the tasks that can be assigned to quality circles through training and increased experience in group problem solving by the members.

### The Pilot Project Approach

A pilot project is a forerunner of a larger project or application with the objective of building confidence in a new idea. A pilot project is a learning experience that demonstrates technical feasibility, costs, benefits, and employee acceptance. Effectively, a "pilot" is a means of planning big and starting small.

Clearly, a pilot project is an approach to experiential learning and to gauging resistance by employees, management, and unions. Some approaches that have been taken are to give the quality circle program to some groups and not others or to implement the concept in all groups and then take it away from some of them. The objective in both cases is to measure productivity, take interviews, and give questionnaires to assess the success of the project.

In many cases, and particularly in the area of quality circle activities, management simply does not know to what extent employees can participate and to what degree they want to. A pilot project is a means of determining valuable input to the planning process without spending an excessive amount of money. It is also a means of finding mistakes and problems early and thereby minimizing exposure. If a pilot project flops, it is chalked up to experience. If a heavily committed project fails, it is blamed on more serious factors.

### Selecting a Climate for Success

The best pilot project for quality circle activities is naturally one with a high probability of success. The object shop or department should contain problems to be solved but not so many that the probability of success is low. It should be assigned to a relatively small close-knit group with good internal communication, so they can aid one another, and an enthusiastic manager. Most importantly, a pilot project must not be placed in a pressure group that may not give a concept a fair evaluation.

The time duration for a pilot project must be long enough for the people to become accustomed to the methods but short enough so the momentum and enthusiastic atmosphere does not subside. Most experts agree that two to six months is the optimum duration.

### Choosing the Right Shop or Department

Since quality circle activities are dependent upon employee contributions, it is necessary to measure people's reaction to the pilot project. Interviews and questionnaires are the best methods of assessment. If a person can achieve personal success through quality circle participation, then that person will remain the best source of support.

## EMPLOYEE ACCEPTANCE

A quality circle program is a change agent, so the success of a project is largely dependent upon the employees' and management's reaction to it. It is important that the people regard the implementation phase as their system. As a result, much of change management is dependent upon an acceptance strategy, resistance management, and proper education and training.

## Acceptance Strategy

A successful acceptance strategy is anticipatory, so fears, resistance, and expectations must be identified early in the implementation plan. The announcement of the plan should be made early on in the change cycle, and significant policy questions have to be addressed. The practice of making a general announcement of a quality circle program, however, and then limiting the initial implementation phase to a few departments can lead to negative results, because enthusiasm can quickly become dampened when expectations are not satisfied. As a result, careful attention should be given to the manner in which pilot projects and early implementation plans are introduced.

The "kick-off" announcement should in general be made to the affected departments by a high-level executive, giving the objectives, benefits to the individual and the organization, expectations, feedback mechanism, and circle implementation schedule.

The acceptance strategy is a link between the strategic and tactical plans of an enterprise and covers personnel policies regarding participation, management support, training, rewards and recognition, management recommendations, and the support of labor groups.

## Resistance Management

Some of the factors that contribute to a reduced resistance are also good management practices in general. Probably the most significant aspect to consider in this area is that the members of a quality circle team will probably not be experienced in group dynamics and therefore will be hesitant to offer views because of the possibility of being "shot down." A latent fear of reprisal for nonparticipation may foster a feeling of coercion in some people. Also, the management presentation of a group's recommendation can easily lead to "no recommendation," if the members fear the process. But the biggest resistance, by far, can come from foremen and supervisors who feel threatened by the process of giving some of the action to employees. Especially cumbersome are foremen who run a "tight ship" and tend to interpret any suggestion as a crack in the "hull." This is one of the reasons it is a good idea to start with the foreman or supervisor as the circle leader. However, it may take considerable coaxing by the facilitator and departmental management to allow the leader's position to be turned over to a member.

### Education and Training

The existence of training programs in modern organizations is presently taken for granted. They are supplied by practically all organizations in every sphere of activity.

An integral part of a quality circle program is training—from the member to the steering committee. Basic to planning, however, is the not-so-obvious fact that three types of learning actually take place. The most widely recognized form is the process of acquiring knowledge and understanding. This type includes gaining ideas, principles, concepts, and facts. Another type of learning is skill acquisition, which may include group participation in quality circle activities as well as work tasks. The last type of learning is the changing of attitudes, goals, interest, and horizons. Only the factual form of learning takes place in sessions held by the leader and facilitator. The second and third types of learning are restricted to participation and quality circle activities.

Fortunately, the learning curve comes into play for all three types of learning, so most participants can contribute substantially after a few training sessions and quality circle meetings.

## SUMMARY

Quality circles are unique in that they combine a participative management technique with a historically structured working environment. The technique is, of course, quality circles implemented as a small-group activity, and the working environment is the classical production shop. The notion of a production shop can encompass workers in manufacturing, banking, the office, and the total domain of the service industry. Because diverse groups can be affected by a quality circle program, planning is particularly significant.

A variety of group plans effectively feed into the total business plan. The systems approach to planning is an attempt to put as much precision into the planning process as possible. Management should insure that the most appropriate concept is applied to the most pertinent problem at the most opportune time. The key dimensions of systems analysis for planning are

- Congruence
- Integrity
- Auditability
- Controllability

A quality circle program should be describable and controllable, and the overall response to business conditions should be predictable.

A *strategy* is the means that management chooses to use an organization's resources to reach its objectives. A strategy is an implicit relationship between the organization, its objectives, its resources, and its environment. A statement of strategy is a mechanism for focusing the attention of management planners on a topic and serves as a communications medium for management review, approval, and support. A strategy statement normally includes the following topics:

- Course of action
- Risks
- Dependencies
- Business environment and conditions
- Resources
- Financial facts
- Alternatives

A strategic planner must be involved with quality circle concepts.

The dimensions of a quality circle strategy are threefold:

- Directions and goals
- A guide for implementation
- A rationale for decision making

The steps in the planning process are well defined and are delineated.

A quality circle program should be a direct response to an explicit organizational need. A quality circle strategy gives three things:

- The current position (Where we are)
- Goals (Where we are going)
- Direction (How we get there)

While the current position and goals are necessary, the direction is a major issue because it serves to establish policies and procedures in the following areas: justification, implementation, employee acceptance, and staffing and organization.

The success of a strategic plan is dependent upon good preparatory work and a sponsor within the organization. Acceptance of a plan is dependent upon justification, which involves both agreement and also a commitment of resources. Methods of financial analysis include the payback method and the cash flow method.

The implementation of a strategic plan for quality circles normally involves a pilot project followed by horizontal and vertical expansion within the organization. Significant factors include choosing the right application and insuring that there is a climate for success. Employee and management acceptance is directly related to resistance management and effective education and training.

# PART THREE
# INDUSTRIAL ROBOTS

# 6 ROBOT PRINCIPLES

## INTRODUCTION

The world's business community has endorsed industrial robotics as one of the key high-tech areas for the 1980s. The reasons are clear:

- Robots are cost effective because wages are rising much faster than robot prices.
- Robots do not decrease the size of the manufacturing work force.
- Robots can be used in hostile environments.
- Robots increase quality and productivity.

Support for these claims is readily available. Unfortunately, the numbers vary, depending upon who is doing the counting. A few numbers, however, can give a visceral feeling for the economics of industrial robotics. First, the average hourly wage of a robot is $4.50, as compared to an $18 hourly labor cost in the automotive industry. Robot prices range from $20,000 to $50,000 with an average payback period of two years. Moreover, robots can do undesirable jobs, such as spray painting, casting, spot welding, and machine loading; also, they survive in unhealthy environments—such as the gaseous hull of a ship during assembly.

The robotics field is booming. The U.S. market alone, reported at $180 million in 1982, is predicted to rise to $2 billion in 1990. In one survey of the number of installed robots, the United States is currently fourth behind Japan, France, and West Germany, in that order. Yet in another study, it is number two behind Japan. It all depends on your definition of a robot, which is covered in the following section. Nevertheless, by 1990, the United States is predicted to be number one. The industry seems to be growing at a rate of approximately 35 percent per year, and the growth rate could easily be higher.

There are approximately 75 firms in the robotics industry in the United States, 200 in Japan, and perhaps 25 in Europe. Japan is reported to build an impressive 45 percent of the world's robots. In the United States, the industry has attracted several large multinationals—including GE, GM, and IBM.

In developed countries, direct human labor runs about 14 to 15 percent of total costs and represents nearly 22 percent of the total work force. By the year 2000, the direct human labor component of the total work force is predicted to drop to 5 percent. The figures are impressive—even though they might turn out to be only partially true.

Many persons have long felt that the lack of machine vision systems has thwarted the growth of robot systems and the fully automated factory. Recent technological advances have produced products that can be applied to parts inspection, product identification, and robot guidance. Roughly 10 percent of the U.S. factory work force—some 600,000 to 700,000 workers—is engaged in inspection and checking operations. It is estimated that from one-third to one-half of these workers could be displaced with vision systems. This market alone totaled $18 million to $19 million in 1982 and is expected to reach $750 million in 1992. A typical vision system costs $30,000 to $35,000 plus $15,000 for special lighting and other facilities.

Thus, there appears to be a promising future for industrial robotics and related systems. So much so, in fact, that at least two cases have been reported in which robots physically assemble products for manufacturing itself in automated factories. In Fujitsu Fanuc's Fuji plant, robots turn out other robots at a rate of 100 per month. In the United States, in Florence, Kentucky, the Yamazaki Machinery Works, Ltd., has opened a fully automated factory that assembles machine tools. The Flexible Manufacturing Factory (FMF), as it is called, will employ only six people, with robots doing the rest of the work.

## ROBOTICS SCENARIO

In most organizations, the decision to use industrial robots is made by top management or a high-level technical committee. The actual implementation process normally involves three steps:

- A search for appropriate applications is made and the most feasible for robot automation are selected.
- An evaluation is made of robot systems that are applicable to the selected applications.
- The applications with the highest probability of success are selected as pilot sites.

It follows that applications that readily take advantage of robot strengths are selected in order to obtain a high probability of success. A learning period is needed for workers and management to gain familiarity with the notion of robotization, and proven mundane applications—such as pellet loading—can serve both economic and educational purposes. Many authors have reiterated the fact that robot installation is much more of a management problem than a technical one.

There are six generally accepted reasons that many senior managers currently favor the use of industrial robots in their organizations:

- The decreased cost of robots and their increased dependability have created an attractive economic package for business.
- There is currently an interest in the quality of work life (QWF), and it is generally accepted that robots will significantly increase it.
- The use of robots in some labor-intensive industries can stem the tide of work flowing to offshore manufacturing.
- The use of robots commonly lowers material costs.
- The large baby boom population in the United States is currently in the work force, but this ample supply of relatively cheap labor will diminish.
- The use of robots contributes to increased productivity and product quality.

In addition to reasons such as these, the acceptance of industrial robots has been accompanied by a closer look at manufacturing technology. What top management sees is a new way of doing business.

Business analysts have determined that manufacturing plants in Japan cost less and use less space than plants producing equal products in the United States. This concept is attributed to a concept known as just-in-time production, which is the heart of effective robotization and total quality control. The objective of just-in-time (JIT) production is to reduce machine setup times so that it is economically feasible to run very small batches. In fact, the ideal goal is to make one piece just in time for the next operation.

This just-in-time concept leads to a characteristic manufacturing modality:

- The JIT modality is to make a piece, check it, and hand it to the next person.

- The batch modality is to produce a high number of parts on a fast machine and then move them somewhere.

With the JIT philosophy, there simply is no place to put the pieces, and by definition, they are passed to the next stage. With the batch modality, the load of pieces is moved by forklift or conveyor to the next operation.

The JIT modality supports *total quality control* (TQC) by exposing trouble spots. When defective pieces are passed to the next stage with JIT production, the second worker tells the first worker about it because he or she also wants to make quota. Thus, problems are resolved promptly and accurately. With batch processing, a load of pieces are produced and used at a later time—perhaps weeks or months. Here, the second worker, upon discovering a defective piece, just throws it aside. All he or she cares is that there are enough good parts to keep busy. A defective rate of as high as 10 percent could be tolerated as simply a deficiency in the productive process. The results are impressive enough to have a unique acronym: JIT/TQC. Another characteristic of JIT production is that errors are caught at the source, in contrast to statistical quality control sampling and inspection after production. The increased awareness of problems, conditions, and causes by workers is essentially what leads to lower rework hours, less material waste, and higher quality associated with JIT/TQC.

The question is actually one of setup time and costs, lot sizes, and inventory cost—even though JIT production has a quality component as well as an inventory component. Thus, there is always a compromise between costs associated with order processing and setup and costs associated with carrying inventory and sustaining less than optimal quality. This compromise is known as *economic order quantity* (EOQ) and is depicted in Figure 6.1. It follows that a major objective of JIT production is to move the EOQ point, as much as possible, towards a value of one.

The move to JIT production has the general objective of eliminating inventory buffers and associated workers. In the latter case, the use of hand tools and production devices ultimately reaches a point of stagnation and workers are replaced by pick-and-place devices and manual manipulators—known as pseudo robots—which are covered later. The inflexibility of pseudo robots subsequently leads to their replacement by programmable robots permitting the flexibility of supporting low setup times and costs.

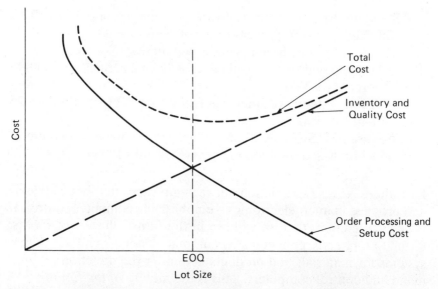

**Figure 6.1** Conceptualization of economic order quantity.

The robotics scenario is essentially complete. Robots are attractive to management because of two key notions:

- Increased productivity in the production process for labor-intensive processing
- Important production in the areas of quality, investor cost, setup time and costs, lot sizes, and manufacturing flexibility

But there is more to the picture. There are a lot of ideas circulating among production people. Some of the key words heard, especially if you listen to vendors, are

- Robot systems
- Sensor-based systems
- Flexible manufacturing
- Production cells

and so forth. As a cautionary note, delphi studies have produced the following predictions (Schreiber, 1982, p. 41):

- The average cost of a robot will decrease from $35,000 (relative to 1980) to $30,000 (relative to 1990).

- Standalone robots, constituting 80 percent of robot sales in 1985, will decrease to 60 percent of robot sales in 1990.
- The use of sensor-based systems will increase.
  — Tactile-based systems will increase from 2 percent (in 1980) to 20 percent (in 1990).
  — Vision systems will increase from 1 percent (in 1980) to 25 percent (in 1990).
- The use of U.S.-made robots will stay the same at 80 percent of units purchased between the years 1980 and 1990.

Last, there is close association between advances in robotic systems and progress in microelectronics—emphasizing computer control. It is likely, therefore, that we will see in the future clusters of robots, termed *robot cells*, connected by automated storage and movement systems for materials and products, all under the direction of command and control computers. This is the *factory of the future*.

## ROBOT CONCEPTS

A robot is an intelligent automation system possessing the following characteristics:

- It possesses memory, control, and decision-making capability.
- It can be trained and can operate under programmed control.
- It can sense the environment to a greater or lesser degree and can respond accordingly.
- It can perform work tasks much like a human worker.

These characteristics are summarized in several carefully worded definitions. The International Standards Organization defines a robot as:

> A machine formed by a mechanism including several degrees of freedom, often having the appearance of several arms ending in a wrist capable of holding a tool or a work piece or an inspection device. In particular its control unit must use a memorizing device, and sometimes it can use sensing and adaptation appliances taking into account environment and circumstances. These multipurpose

machines are generally designed to carry out a repetitive function and can be adapted to other functions.

The Robots Institute of America (RIA) defines a robot as:

> A reprogrammable, multifunctional manipulator designed to move material, parts, tools, or specialized devices through variable programmed motions for the performance of a variety of tasks.

Both definitions are good, but one that is more concise, precise, and technically correct has been offered by Joseph F. Engelberger, known as the father of industrial robotics and the author of a best-selling book on the subject:

> A programmable manipulator with a number of articulations.

What these definitions are saying is that a robot, by American standards, is a general-purpose system that is programmable and flexible.

Flexibility, taken here to mean that a robot is capable of performing a variety of tasks, is not necessarily a requisite for a Japanese robot. As a result, automatons that perform fixed or preset sequences are also regarded as robots in Japan, contributing significantly to the high number that they claim to have in operation.

## ROBOT ANATOMY

An industrial robot is an electromechanical device composed of four major components, as suggested by Figure 6.2:

- Controller
- Power supply
- Manipulator
- End effector

The *controller* is the brain of the operation; the *power supply* is the muscle; the *manipulator* is the arm; and the *end effector* is the wrist and hand. Clearly, not all robots look like Figure 6.2, but the diagram does identify the terms that are used for referring to the components.

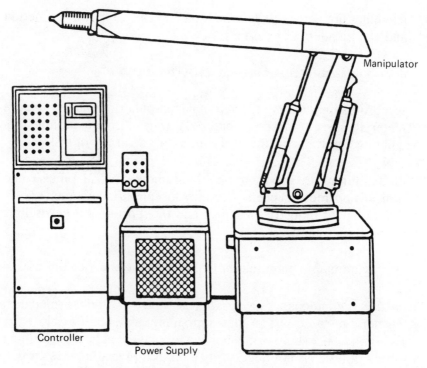

**Figure 6.2** Major components of an industrial robot.

## ROBOT GEOMETRY

A robot is designed to perform a given class of tasks, ranging from minute subassembly to heavy lifting. Each component can be engineered in several ways, and the designer selects the most appropriate modality for an application domain. A key characteristic is the geometry of the manipulator arm and its base, because this movement determines the complexity of the task that can be performed.

In industrial robotics, four basic coordinate systems are used for arm movement:

- Rectangular coordinates
- Cylindrical coordinates
- Spherical coordinates
- Articulated spherical (jointed arm) coordinates

Each of these geometrics is shown in Figure 6.3, and each has a particular class of applications that is best for it. Cylindricals coordi-

ROBOT PRINCIPLES  107

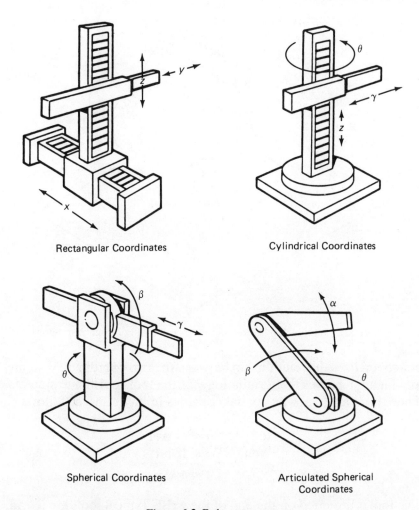

Rectangular Coordinates

Cylindrical Coordinates

Spherical Coordinates

Articulated Spherical Coordinates

**Figure 6.3** Robot geometry.

nates could be best for feeding a machine tool, whereas a jointed arm could be best for reaching inside of a container or for doing electrical assembly. Spherical coordinates have been used for spot welding in the automotive industry.

Each coordinate system provides three degrees of freedom and is capable of placing the end effector (or wrist assembly) in close proximity to the work area. The wrist assembly also has a coordinate system, and the mathematics of robot movement involves the relationship between the two coordinate systems. Manufacturing people call the second coordinate system the *tool coordinate system.* Figure 6.4

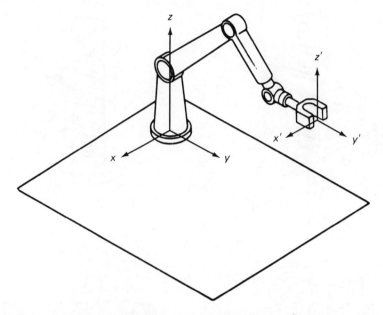

**Figure 6.4** Relationship between the arm and tool coordinate systems.

conceptualizes the relationship between the two coordinate systems, and Figure 6.5 gives the terminology for the tool coordinate systems. There is an analogy with spatial concepts in the terms, as follows:

Roll ↔ Wrist swivel
Pitch ↔ Wrist band
Yaw ↔ Wrist yaw

The tool is mounted on the end of the wrist. Typical tools are a drill, gripper, and spray gun.

The various arm and wrist geometrics are combined to yield a multiplicity of variations ranging from 3 to 6 degrees of freedom. Almost all robots include the three degrees of freedom of the arm assembly, but the wrist assembly can have between one and three degrees of freedom depending upon the application and the tool.

## ROBOT DEVELOPMENT

Industrial robotization has passed through several stages that are more or less related to the current state of development. Each form

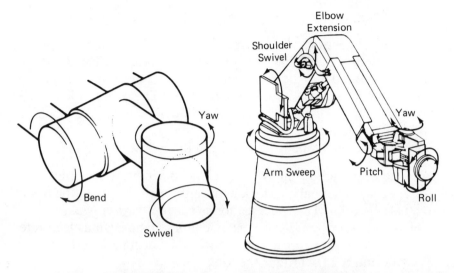

**Figure 6.5** Tool coordinate systems and terminology.

exists today and has features in common with today's robot offerings and tomorrow's prospects.

### Remote Manipulators

The most basic type of industrial automation, related to robotics, is the *remote manipulator arm* (or telecheric, as it is sometimes called). The best-known example of a remote manipulator arm in action is in the nuclear field to handle radioactive materials, where an operator guides the motion of an arm and wrist assembly from a remote location. The remote manipulator arm is viewed through a closed-circuit video system or through thick leaded glass for visual feedback. In fact, a direct mechanical linkage between operator controls and the remote appliance allows the operator to feel the remote operations. In some systems, the operator's force is amplified through servos, allowing an added dimension in performance. A remote manipulator arm is not an industrial robot in any sense of the definition. However, the arm and wrist assembly, linkage system, and servo control of the manipulator arm are similar to those used with robotics. The human operator in a remote manipulation system prevents the system from being classed as a robot because the operation of the device is not automatic.

## Numerical Control

Numerical control is a means of machining metal parts by machines whose actions are described by a sequence of numbers. The sequence of numbers is fed into the control unit of a machine, such as a milling machine or lathe, on a punched paper tape. The operator loads the stock into the machine and aligns the tool. Thereafter, the machine operates automatically.

The data on the tape is developed in the following manner:

- The sequence of steps to machine the part is specified in a special numerical control language through the use of the original blueprint. The specification is called a *part program*.
- The *part program* as well as machining data, such as feed rate and cutter offset, are fed into a program that generates the *center line data* (CLDATA) for the tool.
- The CLDATA is fed into a postprocessor program, specific to a control unit and machine, that generates the input tape.

Numerical control is particularly useful for machining complex parts, such as airframe sections where tolerances and duplicity are of prime importance.

The process of going from a drawing to a part program, CLDATA, paper tape, and control unit is time consuming, and errors occur, requiring that the loop be repeated. A recent advance, known as *direct numerical control,* permits the control unit to accept input data directly from the computer.

As with remote manipulator arms, a numerical control system is not an industrial robot, even though a numerical control system satisfies the requirement for automatic operation. The numerical control machine tool is not a manipulator arm in the sense of an industrial robot, because, as mentioned above, it is a milling machine, lathe, or similar class of machine.

## Pick-and-Place Robots

A *pick-and-place robot* exhibits the most elementary forms of motion through a preprogrammed sequence of steps—each taken to be an end point of a robot limb. Programming a pick-and-place robot is accomplished by setting end points and limit switches to their appropriate positions through an electromechanical pegboard.

The chief characteristic of this type of robot is that the control system does not exercise control over the manipulator arm while it is in motion, so there is no stopping point along a path. Thus, the robot goes by rote between stops. This characteristic tends to make a pick-and-place robot quite noisy; in fact, it has been named the *bang bang machine* for that reason. The motor drives the manipulator arm at full speed until it hits a stop position with a "bang."

Pick-and-place robots are not programmed, but they are set up in a manner quite similar to the way that automatic machines are set up in factories. Setup is time consuming and tedious. Thus, a pick-and-place robot has limited flexibility, even though performance and repeatability of these machines are quite good. The U.S. definition of a robot does not include pick-and-place robots; the Japanese definition does.

**Programmable Robots**

A programmable robot provides the flexibility necessary to conveniently change a robot program in either of two modes:

- A *teach mode,* where an operator takes the manipulator arm through its motions and they are stored in the memory of the control unit
- A *software mode,* where a robot program is written and transferred to the memory of the control unit

In either case, the result is the same. The robot system learns a series of moves that can be repeated upon demand. Moreover, the trajectory of a robot limb is not limited to stops or end points. It can assume any intermediate value that is programmed by the operator.

When a robot of this type operates, it is essentially playing back the sequence that was remembered. Thus, it is often called a *playback robot.*

Playback robots can use point-to-point control or continuous control. With *point-to-point control,* the operator electromechanically moves the manipulator arm in a "teach" mode, stopping at desired points and telling the control unit to remember them. During the playback mode, the robot moves from one memorized point to another. Figure 6.6 gives a representative schematic of robot programming and playback.

## 112 INDUSTRIAL ROBOTS

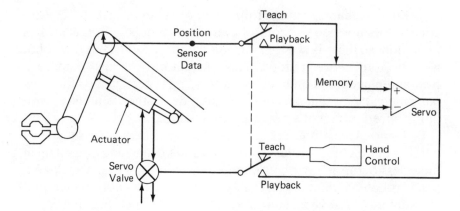

**Figure 6.6** Representative schematic of robot programming and playback.

With *continuous control,* the control unit memorizes a continuous path of points, through sampling, as the operator takes a manipulator arm through its motion in the teach mode. Then, during playback, the exact trajectory is repeated. A typical application of continuous control would involve seam welding.

## ROBOT CONTROL

Robots can also be described by the manner in which they are controlled, and this topic includes actuators, arm control, and memory. Some of the concepts are synonymous with pick-and-place and programmed robots.

### Actuators

The device that moves the manipulator arm is called an *actuator,* and it looks like a piston or shock absorber in some cases. Three ways of powering actuators are used in robotics: pneumatics, hydraulics, and electric motors.

*Pneumatic actuators* are simple and inexpensive but do not give the operational precision required for programmed devices. As a result, pneumatic actuators are primarily used with pick-and-place robots.

*Hydraulic actuators* are used primarily with medium to heavy payloads and are the most frequently used type of actuator. Considerable maintenance is required of actuators of this type, because of oil leaks, so they are primarily used where power is required.

*Electric actuators* are used primarily with small assembly robots. Stepping motors are used to achieve good precision over short distances.

The end effector also requires actuation for object handling and tool manipulation. While open-and-close grippers are widely used as end effectors, the variety of devices is virtually unlimited. Vacuum, mechanical, or magnetic actuators are widely used.

### Arm Control

Stated briefly, robots are nonservo controlled or they are servo controlled. With a *nonservo controller robot,* a typical operating sequence is

- The sequencer selects an axis for movement.
- The sequencer sends a signal to an actuator on the manipulator arm.
- The actuator valves open allowing air to the components.
- The arm moves along the designated axis as long as the valve remains open and the arm is not physically restrained.
- When a physical limit (stop or switch) is reached, the valve closes and the arm stops.
- The sequence goes on to the next step until finished.

Nonservo robots are characterized by high speeds, good repeatability, and low cost. When deceleration is required, valving or shock absorbers are used.

With a *servo controller robot,* the operating sequence is more complex:

- The controller senses the current position of the axis.
- The controller selects a command from memory.
- The data are compared and the controller translates an amplified error signal to a command which is sent to one or more actuators. (More than one axis can move at one time and normally does.)
- Servo valves control pressure to the actuators.
- Feedback devices transmit position back to the controller, which adjusts signals to the actuators.
- When the axes come to rest, the controller addresses the next memory location and the process continues.

A servo-controlled robot is characterized by smooth operation, flexibility in selecting and creating programs, and a higher cost than nonservo devices.

Most robots use microcomputer controllers, and in electric robots, it is not uncommon to have a microprocessor on each axis for control.

## Memory

Whereas earlier programmable robots did not use a random-access memory for points and commands, modern control systems do. This facility has paved the way for robot programs to be prepared off-line in a microcomputer or minicomputer and transferred to the robot's memory through a computer network system. The key advantage of this philosophy is that one microcomputer, for example, can be used to prepare robot programs for a whole shopful of robots.

## ROBOT PROGRAMMING

As introduced previously, there are two ways to program a robot:

- A human operator moves the robot through the desired sequence of motion.
- A robot programming language is used to specify the steps the robot is to execute and the positioning and control data necessary for processing.
- A technician physically sets the stops and switches on a pick-and-place robot. This form is not covered further.

The process of leading a robot through a sequence of steps has two variations: point-to-point control and continuous control. In either case, the operator sequences the manipulator arm with a hand-held box with buttons and switches, called a *teach pendant,* or through a *control console* that serves the same purpose. Figure 6.7 shows a typical teach pendant and control console.

In the case of robot programming languages, the actual programming process normally takes place as follows:

- A robot programmer prepares the sequence of steps and enters them into a "preparation" computer.
- The manipulator arm is moved to key positions and a special

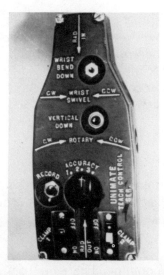

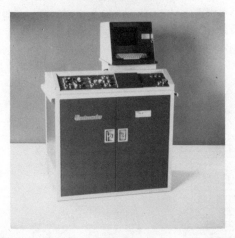

**Figure 6.7** Typical teach pendant *(top)* and control console *(bottom)*. *(Courtesy Unimation)*

operational mode permits the physical data point to be entered into the program directly.
- The robot program is processed by robot software and transferred to the robot's control system.

There is currently a great deal of interest in robot programming languages; two well-done languages in this class are VAL (from Unimation) and AML (from IBM).

## 116 INDUSTRIAL ROBOTS

In preparing a robot program in either of the above modes, the engineer must be aware of the physical area in which the robot can operate. This is called the *work envelopes* and is suggested in Figure 6.8. The work envelope is important for tool and work placement and also for safety. In the latter case, safety barriers are commonly used to isolate the work area.

### SUMMARY

The world's business community has endorsed industrial robotics as one of the key high-tech areas for the 1980s, because robots are cost effective, do not decrease the size of the manufacturing work force, can be used in hostile environments, and increase quality and productivity.

The robotics field is booming. There are over 300 robotics firms worldwide and the industry appears to be growing at a rate of at least 35 percent per year. The use of robots should decrease the size of the direct labor component from 15 percent of total costs to approximately 5 percent. The use of sensor-based systems is also on the upswing. Robots have even been used to build robots in automated factories.

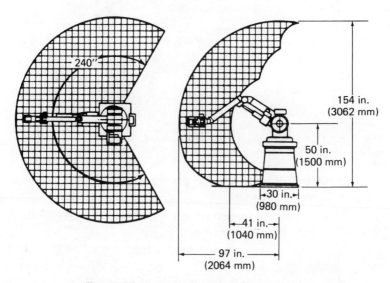

**Figure 6.8** Representative work envelope.

The decision to use industrial robots is usually made by top management or by a high-level technical committee, and promising applications are identified, evaluated, and pilot tested. Pilot selection and training are key issues.

The use of industrial robots goes hand in hand with just in time (JIT) production and total quality control (TQC) to increase productivity, quality, and flexibility, and to enhance inventory control.

It is likely that we will see in the factory of the future clusters of robots, termed robot cells, connected by automatic storage and movement systems for materials and products all under the direction of command and control computers.

A robot is an intelligent automation system with the following characteristics: memory, control, decision-making capability, trainability, sensing capability, and the ability to emulate certain types of work behavior. In general, a robot is a general-purpose system that is both programmable and flexible.

A robot is composed of four major components: controllers, power supply, manipulator arm, and end effector. The four basic coordinate systems used for arm movement are rectangular, cylindrical, spherical, and articulated spherical. Each system has its own advantages for particular applications. In addition to the coordinate system of the manipulator arm, the end effector has an additional system used for tool movement.

The points of reference in robot development are remote manipulators, numerical control, pick-and-place robots, and, finally, programmable robots. Programmable robots can operate in either point-to-point or continuous modes.

A robot is controlled through actuators that can be nonservo or servo driven. Controllers can be pneumatic, hydraulic, or electric, and each has its own problem domain. Pick-and-place robots are nonservo controlled. In modern robots, points and commands are stored in random-access memory. Through systems and software techniques, a single microcomputer or minicomputer can prepare robot programs for several operational robots.

There are three ways to program a robot:

- A human operator moves the robot through the desired sequence of motion.
- A robot programming language is used to specify the steps a

robot is to execute and the positioning and control data necessary for processing.
- A technician physically sets the stops and switches on a pick-and-place robot.

Clearly, the first two methods apply to programmable robots. In the process of taking a robot through the steps (method 1), a teach pendant or control console is used. The second method, of course, uses a robot programming language, closely akin to a computer language. A work envelope is used for application design and safety.

# 7
# ADVANCED TOPICS IN ROBOTICS

**INTRODUCTION**

In the context of productivity, quality circles, and industrial robots, it could be expected that anything but a tried-and-true method would be subjected to close scrutiny. It is true that robot systems are chosen carefully, and dozens of papers exist in professional journals and trade magazines on justification, training, implementation, human relations, employee relations, union relations, and so forth. In spite of an ongoing concern for productivity, however, several advanced concepts are being explored and tested in a commercial environment. This chapter introduces the following important topics:

- Flexible manufacturing systems
- Machine vision systems
- CAD/CAM and robotics

In each case, actual running systems currently exist, but their overall impact on manufacturing has been relatively small.

Robotics is an application-driven field, even though basic research may take place in the areas of artificial intelligence, mathematics, or engineering. The people who actively develop systems are well aware of the direction of robotics and pick and choose research concepts to satisfy needs in productivity and quality.

A brief survey of robot applications yields the topics listed in Table 7.1. An element of commonality exists among the listed applications, and it is that standalone robots could do the job. The greatest benefits to be derived from robotics, however, is when production lines and the design of parts and assemblies are established or modified for the most efficient utilization of robots. Automatic production systems should involve robots as only one type of production element in the system, and not as the only kind or as an end in itself. For example, robots can be used to load and unload numerical control machines

## Table 7.1 Commonly Observed Robot Applications

- Palletizing
- Die casting
- Stamping
- Machine loading/unloading
- Spot welding
- Forging/foundry
- Arc welding
- Plastic molding
- Spray painting
- Assembly
- Cleaning, deburring, polishing

and conveyors and to perform relevant visual inspection and tolerance checking. The biggest payoff occurs, naturally, when three or four machines are arranged so they can be serviced by one robot. Clearly, the best results are obtained when worker intervention is kept to a minimum.

These comments provide the operational basis for the topics covered.

## FLEXIBLE MANUFACTURING SYSTEMS

The just-in-time/total quality control (JIT/TQC) concept presented in Chapter 6 is a desirable modality provided that the production process can support it. The strengths of JIT/TQC are by-products of flexible manufacturing systems.

A *flexible manufacturing system* (FMS) is a production system that can be programmed to alter its procedures in response to varying production requirements. An FMS should be contrasted to a hard automation system that adheres to a fixed sequence of steps in manufacturing a product. An FMS will probably contain one or more robot cells (or work cells), but is not limited to the cell concept.

A flexible manufacturing system permits a company to produce multiple parts and multiple assemblies through programming. With FMS, small batches and JIT production are feasible. Batch processing is labor intensive and start-up costs are high. Even though production volume is good with batch systems, it has been estimated that parts and assembled inventory spends 90 percent of its time waiting. Another problem with batch processing is that setup is so rigid that previously produced components may not conform to the new setup. With an FMS, the production process can easily be adjusted, thereby reducing rework and scrap.

One manufacturing executive is quoted as saying, "Flexible systems make possible economics of scope as well as scale." In short, flexible manufacturing permits the automation of low-volume production. Heretofore, automation has been the province of high-volume runs.

Another aspect of flexible manufacturing systems is the inspection process. Statistical sampling techniques are fine for batch manufacturing, since a process stabilizes once it is honed in. When a setup is constantly changing, statistical techniques are not applicable and an FMS should inspect every part.

The result has been the inclusion in a fabrication, machining, or assembly system of an inspection station. Based on laser technology and flexible vision systems, coordinate measurement and detection systems now provide automatic inspection as well as production.

In a fully automated system, in-process failures can occur to tools and machines. This is another aspect of an FMS, even though a flexible system can live without it.

Many experts now feel that an automated factory in general—and a flexible manufacturing system in particular—needs more than process automation. New control modes are predicted. A *hierarchical control system* for manufacturing would analyze a manufacturing task, subdivide it into subassemblies, and assign each subassembly to a manufacturing cell. This concept, which is directly related to rule-based programming in knowledge-based systems, requires even more than flexible manufacturing. It requires a "world view" on the part of the control system and new concepts in plant design and utilization.

Many specialists see a need for plant modularity wherein manufacturing cells can be added (or deleted) to the total system configuration on a dynamic basis. Thus, manufacturing cell 1 would be assigned to task A for some period and then be reconfigured to task B.

Information systems also play a key role in flexible manufacturing. Analyses must be performed of the salient characteristics of parts and assemblies so that items with similar manufacturing characteristics can be grouped and be produced by the same FMS.

## MACHINE VISION SYSTEMS

Critics of the modern state of robotics base their claims on the fact that robots do little to sense their environment and, as such, are essentially numerically controlled machines. In order to exhibit intelligent behavior, they say, robots must monitor external conditions and events and respond accordingly. Two methods of sensing the environment are considered to be most useful for robotics: tactile sensing and machine vision. Of the two possibilities, machine vision is considerably far ahead of tactile in technological development and applications.

Machine vision is particularly useful in robotics because it enlarges the scope of tasks that can be performed. Moreover, it just so happens that the jobs that management, workers, and unions would like to turn over to robots are the ones that require visual capability. Jobs in this class are dangerous and lend themselves to boredom and fatigue. In some cases, clearly related to the environment, machine vision systems perform better than their human counterparts.

Representative examples of machine vision in industrial robotics are inspection, part and assembly identification, bin picking, and machine guidance and control. The task of picking up a randomly oriented piece jumbled in a bin was a tough nut to crack, but several vendors have done it. In fact, the difficulty led several engineers to claim that bin picking was not the right problem, but rather, reorientation of the work flow and environment was the key issue. It all sounds like JIT/TQC production.

Most machine vision systems can recognize silhouette imagery when there is sufficient contrast between the object and background. Difficulties arise when objects have different colors or textures. Actually, these are even problems in some cases for human beings. Recent techniques, associated with three-dimensional machine vision, measure light intensity and are thereby able to distinguish between colors and textures.

Machine vision systems are classified according to how they receive, translate, and process images. In *two-dimensional technology,* either a binary scale or a gray scale is used. In a *binary system,* each pixel in an image is translated into either black or white through a thresholding technique. In a *gray system,* varying shades of gray are recorded. The binary system is sufficient for many computer applications, as is the gray system, which provides more information to the system because of the varying intensities. Actually, neither is particularly useful for robotics, per se, but can serve a useful purpose in inspection and identification.

Three-dimensional machine vision technology is capable of sensing depth and can assist a robotic system in maneuvering in space. Considerable machine reasoning ability, however, is needed to fully utilize this capability.

Three basic methods of machine vision have been developed for industrial robotics:

- Range finding
- Structured light
- Binocular vision

In a *range finding system,* the time needed to reflect a laser beam off the object is measured and used to compute its distance. In a *structured light system,* a controlled light is projected on the object and its distance is measured by triangulation. With *binocular vision,* an attempt is made to emulate human vision by using two cameras, creating a binocular parallax.

The basic method of recognition in machine vision is through the use of exemplars. In fact, this technique permits partial images to be recognized.

In the operation of a machine vision system, an electronic retina from the camera system is scanned and sampled and an image is formed in a random-access memory in the form of pixels. Each pixel may require one or two bytes of data, depending upon the intensity levels. For long-term or temporary storage, an image can be placed on magnetic disk. A conceptual machine vision system is given in Figure 7.1. A machine vision can use the following components:

- *Camera* for image generation
- *Photocell* for triggering purposes
- *Strobe light* for "stopping" the action
- *Vision computer* for image processing
- *Video monitor* for monitoring the image analysis
- *Keyboard* for controlling the vision computer
- *Host computer* for large-scale processing
- *Robot system,* which is the control object

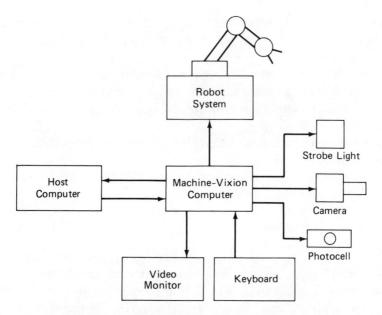

**Figure 7.1** Conceptual machine vision system.

Image analysis is extremely sophisticated and normally uses two levels of computation: image preprocessing in the vision computer and "heavy duty" computing in a scientific (host) computer.

The mathematical algorithms for image analysis and recognition constitute a specialty, per se, with the field of machine vision. Normally, different methods apply to binary systems, gray systems, and three-dimensional systems.

In spite of the complexity of machine vision systems, the field is booming. A representative vision module, designed to supplement a robot system, runs from $15,000 to $40,000 and adds about 30 to 50 percent to the price of the robot system. Industry observers predict that the percentage should drop to approximately 15 percent as the industry matures. Approximately 20 machine vision firms currently exist in the United States (Kinnucan, April, 1983), and the field is expected to grow.

## CAD/CAM AND ROBOTICS

Computer-aided design/computer-aided manufacturing (CAD/CAM) is the use of specialized hardware and software to facilitate the design and manufacturing operations. It is important to recognize that computer-aided design (CAD) is not generally regarded as traditional scientific computation and computer-aided manufacturing (CAM) is not MIS. Table 7.2 lists a representative sample of CAD/CAM systems.

Computer-aided design is more of a process than it is a system. CAD, as a discipline, is engineering oriented and concerns design, analysis, administration, and information management relative to engineering functions. Thus, the purpose of a CAD system is to allow the engineer and designer to interact with a computer using graphic techniques, special languages and systems, database facilities, and special computation hardware and software. In the design process, the designer can create, modify, save, and produce hard copy from an interactive video display unit. In the engineering process, the engineer can perform computations and engineering analyses using special software packages on designs produced via the interactive graphics facilities. Computer-aided design provides considerably more to engineering management than a means of increasing the productivity of designers and engineers. CAD, as a discipline, provides a means of achieving consistency in the design and engineering processes and a

**Table 7.2 Representative List of CAD/CAM Systems[a]**

| Function | Examples |
| --- | --- |
| Design layout | Computer-aided drafting, printed circuit board layout |
| Design analysis | Computer optimization, finite element analysis, piping interference checking |
| Manufacturing engineering | Group technology, tool design, process planning |
| Facilities engineering | Plant architecture and layout, equipment optimization |
| Fabrication automation | Numerically controlled tools, process control systems |
| Assembly automation | Robotics, computer-controlled transfer lines |
| Materials handling | Stacker cranes, driverless tractor systems, automatic storage and retrieval systems |
| Industrial engineering | Shop floor data collection, labor standards calculations |
| Quality assurance | Coordinate-measuring machines, automated circuit test equipment |

[a]Reprinted with permission from *Encyclopedia of Computer Science and Engineering,* Copyright 1983 by the Van Nostrand Reinhold Company.

mechanism for achieving an element of control over creative work based on technical concepts.

Computer-aided manufacturing is a broad class of functions involving manufacturing operations, such as fabrication, assembly, and inspection, and involving manufacturing control, such as scheduling, inventory, and requirement planning. Thus, the term *computer-aided manufacturing* is more of a shell for a collection of related technologies, such as:

- NC—numerical control
- MRP—manufacturing resource planning
- CPM—critical path method
- Industrial robotics

than it is unified technology.

An important point to be recognized is that the acceptance by management of industrial robotics and numerical control as key elements in the manufacturing process has generated requirements that parts and assemblies be designed and engineered for manufacturing, as well as for use. This has created a strong bridge between CAD and CAM—hence the name CAD/CAM.

## SUMMARY

Because of the ongoing concern of management over productivity and quality, it would be expected that any advanced topic in industrial robotics would be subject to close scrutiny. While it is in fact true that dozens of papers have been written on justification, training, employee relations, and so forth, several advanced concepts have surfaced that deserve a brief introduction. Three of these topics are introduced here:

- Flexible manufacturing systems
- Machine vision systems
- CAD/CAM and robotics

Each of these areas has been explored technically and is now being tested in commercial environments.

The greatest benefits are derived from robotics when production lines and the design of parts and assemblies are established or modified for the most efficient utilization of robots. Automatic production systems should involve robots as only one type of production element in the system, and not as the only kind or as an end in itself.

A flexible manufacturing system (FMS) is a production system that can be programmed to alter its procedures in response to varying production requirements. An FMS should be contrasted to a hard automation system that adheres to a fixed sequence of steps in manufacturing a product. An FMS will probably contain one or more robot cells (or work cells) but is not limited to the cell concept. Just-in-time/total quality control (JIT/TQC) are by-products of a flexible manufacturing system.

A flexible manufacturing system permits a company to produce multiple parts and multiple assemblies through programming. With FMS, small batches and JIT production are feasible, and the production process can easily be adjusted, thereby reducing rework and scrap. Based on laser technology and flexible vision systems, coordinate measurement and detection systems now provide automatic inspection as well as production. Three advanced topics associated with flexible manufacturing systems are hierarchical control systems, modular manufacturing cells, and manufacturing information systems.

In order to exhibit intelligent behavior, robots must monitor external conditions and events and respond accordingly. Machine vision systems are the most developed methods of sensing the environment. Machine vision is particularly useful in robotics because it enlarges the range of tasks that can be performed. Representative applications of machine vision in industrial robotics are inspection, part and assembly identification, bin picking, and machine guidance and control.

Machine vision systems are classified according to how they receive, translate, and process images. The basic method of recognition in machine vision is through the use of examplars. The machine vision field is booming and at least 20 firms in the United States specialize in this area. The inclusion of a machine vision system currently adds 30 to 50 percent to the price of a robot system. This percentage is expected to drop to approximately 15 percent as the field matures.

Computer-aided design/computer-aided manufacturing (CAD/CAM) is the use of specialized hardware and software to facilitate the design and manufacturing operations. Computer-aided design is more of a process than it is a system. As a discipline, CAD is engineering oriented and concerns design, analysis, administration, and information management relative to engineering functions. The purpose of a CAD system is to allow the engineer and designer to interact with a computer using graphics techniques, special languages and systems, database facilities, and special computer hardware and software.

Computer-aided manufacturing is a broad class of functions involving manufacturing operations and control. Manufacturing operations include fabrication, assembly, and inspection. Manufacturing control includes scheduling, inventory, and requirements planning. The concept of computer-aided manufacturing is a shell for a collection of related technologies, such as numerical control, manufacturing resource planning, critical path method, and industrial robotics.

Through numerical control and industrial robotics, a strong bridge has been constructed between CAD and CAM—hence the name CAD/CAM.

# REFERENCES

Acebes, C., "Companies Pour into Robot Market," *High Technology* (July, 1983), p. 75.
Albus, J. S., *Brains, Behavior, and Robotics,* Peterborough, N.H.: BYTE Books, 1981.
Amsden, R. T., and Amsden, D. M., "A Look at QC Circles," in *Quality Assurance: Methods, Management, and Motivation* (H. J. Bajaria, Editor), Dearborn, Mich.: Society of Manufacturing Engineers, 1981, pp. 225-230.
Bajaria, H. J., "Methods of Quality Control," in *Quality Assurance: Methods, Management, and Motivation* (H. J. Bajaria, Editor), Dearborn, Mich.: Society of Manufacturing Engineers, 1981.
Bajaria, H. J. (Editor), *Quality Assurance: Methods, Management, and Motivation,* Dearborn, Mich.: Society of Manufacturing Engineers, 1981.
Beardsley, J. F., *Quality Circles,* Midwest City, Okla.: International Association of Quality Circles, 1982.
Buehler, V. M., and Shetty, Y. K., *Productivity Improvement: Case Studies of Proven Practice,* New York: AMACOM, 1981.
Callahan, J. M., "The State of Industrial Robots," *Byte* (October, 1982), pp. 128-142.
Crosby, P. G., *Quality is Free,* New York: McGraw-Hill, 1979.
Davis, C. R., "A Strategy for Improving Product Quality Through Quality Awareness and Participative Management," in *Quality Assurance: Methods, Management, and Motivation* (H. J. Bajaria, Editor), Dearborn, Mich.: Society of Manufacturing Engineers, 1981, pp. 17-26.
Demers, K. P., and Walsh, P. M., "Sensor-based Real-Time Robot Control Systems," *Robotics Today* (June, 1983), pp. 69-72.
Dewar, D. L., *The Quality Circle Guide to Participative Management,* Englewood Cliffs, N.J.: Prentice-Hall, 1980.
Drucker, P., *Managing in Turbulent Times,* London: Pan Books, 1980.
Engelberger, J. F., *Robotics and CAD/CAM,* Dearborn, Mich.: Society of Manufacturing Engineers, 1977.
Engelberger, J. F., *Robots in Practice: Management and Applications of Industrial Robots,* New York: American Management Associations (AMACOM), 1980.
"Eyes for the Automated Factory," *High Technology* (April, 1983), p. 37.
Fishlock, D., "BL Plans to Use More Robots," London: *Financial Times* (February 20, 1982).
Freund, R. A., "The Role of Quality Technology," in *Quality Assurance: Methods,*

# REFERENCES

*Management, and Motivation* (H. J. Bajaria, Editor), Dearborn, Mich.: Society of Manufacturing Engineers, 1981, pp. 10-13.

Froelich, L., "Robots to the Rescue?" *Datamation* (January, 1981), pp. 85-104.

Gevarter, W. B., "Robotics: An Overview," *Computers in Mechanical Engineering* (August, 1982), pp. 43-49.

Glaser, R., and Glaser, C., *Managing by Design,* Reading, Mass.: Addison-Wesley, 1981.

Glorioso, R. M., and Osorio, F. C. C., *Engineering Intelligent Systems: Concepts, Theory, and Applications,* Bedford, Mass.: Digital Press, 1980.

Hall, D. M., *Management of Human Systems,* Cleveland: Association for Systems Management, 1971.

Hapgood, F., "Inside a Robotics Lab: The Quest for Automatic Touch," *Technology Illustrated* (April, 1983), pp. 19-22.

Hapgood, F., "Inside a Robotics Lab: Avoiding Obstacles," *Technology Illustrated* (May, 1983), pp. 33-35.

Hapgood, F., "Inside a Robotics Lab: Looking for Stereo Vision," *Technology Illustrated* (June, 1983), pp. 44-49.

Ingle, S., *Quality Circles Master Guide: Increasing Productivity with People Power,* Englewood Cliffs, N.J.: Prentice-Hall, 1982.

*International Conference on Quality Control Proceedings,* Tokyo: Union of Japanese Scientists and Engineers, 1978.

Ishikawa, K., *Guide to Quality Control,* Tokyo: Asian Productivity Organization, 1976.

Ishikawa, K., *QC Circle Koryo: General Principles of the QC Circle,* Tokyo: Union of Japanese Scientists and Engineers (JUSE), 1980.

"Japan's High-Tech Challenge," *Newsweek* (August 9, 1982), pp. 48-59.

Jarvis, R. A., "A Computer Vision and Robotics Laboratory," *Computer* (June, 1982), pp. 8-22.

Juran, J. M., and Frank M. Gyrna, Jr., *Quality Planning and Analysis,* New Delhi: Tata McGraw-Hill Publishing Company, Ltd., 1970.

Kendrick, J. W., "Background and Overview of Productivity Improvement Programs," in *Productivity Improvement: Case Studies of Proven Practice,* New York: AMACOM, 1981, pp. 14-27.

Kinnucan, P., "How Smart Robots Are Becoming Smarter," *High Technology* (September/October, 1982), pp. 32-40.

Kinnucan, P., "Machines That See," *High Technology* (April, 1983), pp. 30-36.

Kinnucan, P., "Flexible Systems Invade the Factory," *High Technology* (July, 1983), pp. 32-43.

Krasnoff, B., *Robots: Reel to Real,* New York: ARCO Publishing, Inc., 1982.

Lipinski, T. E., and Skinner, C. S., "Is There a Robot in Your Future?" *Computerworld* (n.d.), pp. 23-30.

Macarov, D., *Worker Productivity: Myths and Reality,* Beverly Hills: Sage Library of Social Research, 1982.

McGregor, D., *The Human Side of Enterprise,* New York: McGraw-Hill, 1960.

*Microelectronics, Productivity, and Employment,* Paris: Organization for Economic Co-Operation and Development, 1981.

Moritani, M., *Japanese Technology,* Tokyo: The Simul Press, Inc., 1982.
Norman, C., *Microelectronics at Work: Productivity and Jobs in World Economy,* Washington: Worldwatch Institute, 1980.
Nowak, G., "The Advent of Machine Vision Systems," *Manufacturing Engineering* (November, 1982), pp. 56-60.
Ohmae, K., *The Mind of the Strategist,* New York: McGraw-Hill, 1982.
Puri, S. C., and McWhinnie, J. R., "Quality Management Through Quality Indicators: A New Approach," in *Quality Assurance: Methods, Management, and Motivation,* (H. J. Bajaria, Editor), Dearborn, Mich.: Society of Manufacturing Engineers, 1981, pp. 34-40.
Raibert, M. H., and Sutherland, I. E., "Machines That Walk," *Scientific American* (January, 1983), pp. 44-53.
Ralston, A., and Reilly, E. D., Jr. (Editors), *Encyclopedia of Computer Science and Engineering* (2nd edition), New York: Van Nostrand Reinhold, 1983.
"Robots Get a Warm Welcome in Kentucky," *Business Week* (April 19, 1982), p. 30.
Schonberger, R. J., *Japanese Manufacturing Techniques,* New York: The Free Press, 1982.
Schreiber, R. R., "Robotics in the Eighties," *Robotics Today* (October, 1982), pp. 39-42.
Schreiber, R. R., "Robot Vision: An Eye to the Future," *Robotics Today* (June, 1983), pp. 53-57.
Schumacher, E. F., *Good Work,* London: Jonathan Cape, 1979.
Simons, G. L., *Robots in Industry,* Manchester, England: NCC Publications, 1980.
Stauffer, R. N., "Programs in Tactile Sensor Development," *Robotics Today* (June, 1983), pp. 43-49.
Steiner, G. A., *Strategic Planning: What Every Manager Should Know,* New York: The Free Press, 1979.
Sullivan, M. J., "The Right Job for Robots," *Manufacturing Engineering* (November, 1982), pp. 51-60.
"Surveys Reveal Robot Population and Trends," *Robotics Today* (February, 1982), pp. 79-80.
Tanner, W. R., *Basics of Robotics,* Dearborn, Mich.: Society of Manufacturing Engineers, 1977.
Tanner, W. R. (Editor), *Industrial Robots: Volume 1/Fundamentals* (2nd edition), Dearborn, Mich.: Society of Manufacturing Engineers, 1981.
Tarvin, R. L., *Design of a Computer-Controlled Industrial Robot for Maximum Application Flexibility,* Cincinnati, Ohio: Cincinnati Milacron, 1980.
"The Push for Dominance in Robotics Gains Momentum," *Business Week* (December 14, 1981), pp. 108-109.
"The Robot Boom," *Popular Computing* (April, 1982), p. 14.
Thomas, E. F., "Shortcomings of Current Motivational Techniques," in *Quality Assurance: Methods, Management, and Motivation* (H. J. Bajaria, Editor), Dearborn, Mich.: Society of Manufacturing Engineers, 1981, pp. 87-96.
Toffler, A., *The Third Wave,* New York: Bantam Books, 1980.
Twiss, B. C. (Editor), *The Managerial Implications of Microelectronics,* London: The Macmillan Press, 1981.

Veen, B., "Integration of TQC and Motivation Programs," in *Quality Assurance: Methods, Management, and Motivation,* (H. J. Bajaria, Editor), Dearborn, Mich.: Society of Manufacturing Engineers, 1981, pp. 97-101.

Waldman, H., "Update on Research in Vision," *Robotics Today* (June, 1983), pp. 63-67.

Wilensky, R., *Planning and Understanding: A Computational Approach to Human Reasoning,* Reading, Mass.: Addison-Wesley, 1983.

# INDEX

Absence of quality, 12
Acceptance criteria, 17
Acceptance or rejection, 22
Acceptance strategy, 92
Accuracy, 24
Actuators, 112, 117
Applicative, 79, 81
Arm control, 113
Articulated spherical coordinates, 106, 107
Attitude indicators, 64
Auditability, 83, 84, 94
Automation, 4, 10
Average, 24
Awareness teams, 27

Bang bang machine, 111
Bartering, 4
Basic cause and effect, 77, 78, 81
Batch modality, 102
Beardsley, J. F., 44, 46
Binary system, 122
Binocular vision, 123
Bottom line, 11
Brainstorming, 65, 80
   analysis of ideas, 67
   guidelines, 66

CAD/CAM, *see* Computer aided design/computer aided manufacturing.
Cash flow method, 90, 95
Cause and effect diagram, 77, 81
Cause problem, 54
Center line data, 110
Cheap labor, 101
Check sheet, 69
Checklist, 69
Choosing the right shop or department, 91, 95

Circle leader, 44, 50, 57
   functions, 52
Circle member, 44, 57
   functions, 53
CLDATA, *see* Center line data.
Climate for success, 91, 95
Code of conduct, 53
Communications, 40
Competitiveness, 6
Computer aided design/computer aided manufacturing, 119, 124, 126, 127
Concept phase, 18, 31
Conformance analysis, 22, 31
Conformance to design objectives, 17
Congruence, 83, 94
Consistency, 10
Contents of the strategy, 86, 94
Continuous control, 112, 117
Control chart, 71, 81
Control console, 114, 118
Controllability, 83, 84, 94
Controller, 105, 117
Cooperative behavior, 41
Coordinator, 38
Correlation, 73
Cost analysis, 62, 80
Cost of quality, 13, 77
Cost reduction, 62, 80
Cost savings, 90
Current position, 86, 95
Cylindrical coordinates, 106, 107

Data analysis, 74, 81
Data processing, 11
Data recording, 22, 31
Decreased cost of robots, 101
Design change phase, 21, 31
Design phase, 20, 31
Design review, 22

# INDEX

Developing the plan, 87
Direct numerical control, 110
Direction, 86, 87, 94, 95
Division of labor, 5, 6

Economic order quantity, 102
Education and training, 93
Efficient operation, 41
Effort, 10
Electric actuators, 113, 117
Employee acceptance, 91, 95
End effector, 105, 117
Exemplars, 123, 127

Facilitator, 36, 37, 38, 44, 39, 57
    functions, 50
Factory of the future, 104, 117
Feasibility study phase, 19, 31
Fitness for use, 17
Flexible manufacturing system, 119, 120, 126, 127
FMS, *see* Flexible manufacturing system.
Formal leader, 36, 57

Goals, 86, 94, 95
Good work, 10
Grade of product or service, 17
Gray system, 122
Guidelines, 88
Guns and butter diagram, 8

Hierarchical control system, 121
Histogram, 72, 81
Human operator, 111, 114, 117
Hydraulic actuators, 112, 117

Implementation, 90, 95
Improved cash flow, 62, 80
Increased productivity, 101
Ingle, S., 44, 45
Inspection, 22, 31
Integrity, 10, 83, 84, 94
Intellectual domain, 10
Inventory component, 102

Job design, 13
Job satisfaction, 3
Jobs, 10
    white collar, 11
Joint quality circles, 67
Jointed arm coordinates, 106
Just-in-time production, 101, 117, 120, 127
Just-in-time modality, 101
Justification, 88

Kinnucan, P., 124

Learning curve, 62, 80
Line graph, 70, 81

Machine vision, 119, 122, 126, 127
Management development/organization
    development, 37, 60
Management problem, 101
Management steering committee, 36
Manipulator, 105, 117
Manual manipulators, 102
Manufacturing plan, 22
Material cost, 101
Measurement, 60, 80
Mechanization, 4, 10
Meet regularly, 36, 57
Memory, 114, 117
Metaplan, 85
Methodology, 89
Methods of analysis, 89
Methods of quality control, 22-25
Minicircles, 67
Money, 4
Morale, 42

New technology, 15
Nonservo controller robot, 113, 117
Numerical control, 110

Office automation, 11
Offshore manufacturing, 101
Open communications, 14
Operation of a quality circle, 53-55
Ownership, 9

# INDEX

Pareto diagram, 23, 31, 74, 81
Part program, 110
Participation, 36, 57
Participative management, 26, 82, 93
Payback method, 62, 80, 90, 95
Physical domain, 10
Pick-and-place robots, 102, 110, 117, 118
Pictogram, 73
Pie chart, 73, 81
Pilot project, 60, 80, 90, 95
Pitch, 108
Playback robot, 111
Pneumatic actuators, 112, 117
Point-to-point control, 111, 117
Policy committee, 36
Policy formulation, 39
Policy set, 42, 43
Policy structure, 42
Postprocessor program, 110
Power supply, 105, 117
Precision, 24
Presentation of collected data, 70
Process cause and effect, 77, 79, 81
Product design, 13
Product evaluation, 22, 31
Production phase, 20, 31
Productive function, 8
Productive system, 6-8
Productivity, 3-9, 12, 14, 40, 89
   concept of, 4
   definition of, 4
   importance of, 5
Productivity gain, 62, 80
Productivity improvement, 14
Productivity improvement program, 13-14
Productivity objectives, 14
Profitability, 12
Programmable robots, 111, 118
Project chart, 73
Prototype phase, 20, 31
Pseudo robots, 102
Punishment, 4

Quality, 7, 16-32, 40
   definition of, 17, 18, 30
   significance, 15

Quality assurance, 21, 31, 60, 69
Quality assurance plan, 21, 31
Quality awareness, 25-27, 31
Quality circle, 35ff, 57
   methods, 60
   operation, 53-55
   varieties of, 55-56
Quality circle facilitator, 36, 37, 38, 44, 49, 57
Quality circles, 20, 27
Quality component, 102
Quality indicator, 61, 63, 80
Quality life cycle, 18, 31
Quality objectives, 27-29, 32
Quality of work life, 101
Quality organization, 29-30, 32
Quality planning, 27-29
Quality policy, 27-29, 32

Range finding system, 123
Recording of collected data, 68
Rectangular coordinates, 106, 107
Reliability and maintainability, 17
Remote manipulator arm, 109, 117
Resistance management, 92
Result problem, 54
Right shop or department, 91, 95
Robot, 4, 104, 117
Robot anatomy, 105, 117
Robot cell, 104, 117
Robot control, 112, 117
Robot development, 108
Robot geometry, 106, 117
Robot programming, 114, 117
Robot programming language, 115, 117
Roll, 108

Savings-to-cost ratio, 62
Scatter diagram, 73
Schreiber, R. R., 103
Servo controller robot, 113, 117
Similar work, 36, 57
Small group, 36, 57
Small group activity, 82, 93
Software mode, 111
Specialization, 5, 6

Specification document, 22, 31
Spherical coordinates, 106, 107
Sponsor, 87
Statement of strategy, 85, 94
Steering committee, 36, 44, 47, 57
   functions, 48
Strategic plan, 87
Strategic planning, 15
Strategy, 84, 94
Stratification, 73
Structured light system, 123
Structured working environment, 82, 93
Study groups, 67
Suitability for intended purpose, 17
Success elements, 56
Sweatshops, 4

Table pounder, 48
Tactile sensing, 122
Teach mode, 111
Teach pendant, 114, 118
Team approach, 22
Telecommunications, 11
Theory X management style, 35
Theory Y management style, 35
Three-dimensional machine vision, 123
Tool coordinate system, 107, 117
Top down approach, 60
Top-down implementation, 45, 46

Top management involvement, 13
Total business plan, 82, 83
Total organization, 15
Total productivity, 3
Total quality control, 16, 102, 117, 120, 127
Triangulation, 123
Two-dimensional technology, 122

Unionization, 5, 6
Utilizing the plan, 88

Voluntary participation, 36, 57

White-collar jobs, 11
Work, 9-10, 15
   emotional needs for, 9
   inputs to, 9
   output of, 9
   social needs for, 9
Work envelope, 116, 118
Work force, 101
Worker involvement, 14
Worker specialization, 5
Working smarter, 6, 37
World view, 121

Yaw, 108

Zero defects, 20, 27